Relativistic Mechanics, Time and Inertia

Fundamental Theories of Physics

A New International Book Series on The Fundamental Theories of Physics: Their Clarification, Development and Application

Relativistic Mechanics, Time and Inertia

by

Emil Tocaci

Institute of Civil Engineering, Bucharest, Romania
Institute of Oil and Gas, Ploesti, Romania

Edited and with a Foreword by C. W. Kilmister

D. Reidel Publishing Company

A MEMBER OF THE KLUWER ACADEMIC PUBLISHERS GROUP

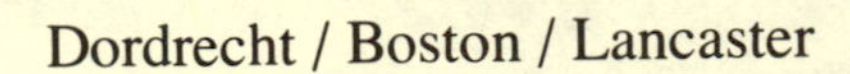

Dordrecht / Boston / Lancaster

Library of Congress Cataloging in Publication Data

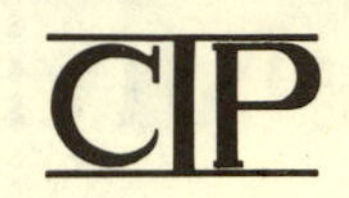

Tocaci, Emil.
Relativistic mechanics, time, and inertia.

(Fundamental theories of physics)
Translation with revisions of: Mecanica relativistă, timpul şi inerţia.
Includes index.
1. Relativistic mechanics. 2. Time. 3. Inertia (Mechanics)
I. Kilmister, C. W. (Clive William) II. Title. III. Series.
QA808.5.T6713 1984 530.1'1 84-9763
ISBN 90-277-1769-9

Published by D. Reidel Publishing Company
P.O. Box 17, 3300 AA Dordrecht, Holland.

Sold and distributed in the U.S.A. and Canada
by Kluwer Academic Publishers,
190 Old Derby Street, Hingham, MA 02043, U.S.A.

In all other countries, sold and distributed
by Kluwer Academic Publishers Group,
P.O. Box 322, 3300 AH Dordrecht, Holland.

Translated from the Romanian by Vasile Vasilescu.

This book is an extensively revised version of *Mecanica Relativista, Timpul si Inertia*, published in 1980 by Editura Stiintifica si Enciclopedica, Bucharest.

Printed in The Netherlands

TABLE OF CONTENTS

FOREWORD

To accept the special theory of relativity has, it is universally agreed, consequences for our philosophical views about space and time. Indeed some have found these consequences so distasteful that they have refused to accept special relativity, despite its many satisfactory empirical results, and so they have been forced to try to account for these results in alternative ways. But it is surprising that there is much less agreement about exactly what the philosophical consequences are, especially when looked at in detail. Partly this arises because the results of the theory are derived in an elegant mathematical notation which can conceal as much as it reveals, and which, accordingly, offers no incentive to engage in the thankless task of dissection.

The present book is an essay in careful analysis of special relativity and the concepts of space and time that it employs. Those who are familiar with the theory will find here (almost) all the formulae with which they are familiar; but in many cases the interpretations given to the terms in these formulae will surprise them. I doubt if this is the last word about these interpretations: but I believe that the book is valuable in drawing attention to the possibility of more open discussion in general, and in particular to the fact that acceptance of the theory of relativity need not commit one to every detail of conventional interpretation of its terms.

King's College London C. W. KILMISTER

A. ASPECTS OF RELATIVISTIC MECHANICS

A.I. GENERAL REMARKS

The essence of fundamental notions has always challenged scientific and philosophical thinking. There is a spiritual striving for an ever more profound understanding of nature that is part of the natural thirst for knowledge. Moreover, the evolution of science is directly related to industrial development, and this is a prerequisite of the fulfillment of the numerous needs of mankind.

The reflection of the universe in human consciousness has evolved through inevitable successive approximations. These approximations have added up and determined qualitative jumps in the elaboration and formulation of the laws of nature. The theory of relativity was such a jump and has become one of the main components of our scientific inheritance.

The very brief outline of the basic ideas of the theory of relativity that follows serves as an introduction to the core of this book.

A.II. REMARKS ON THE THEORY OF RELATIVITY

The theory of relativity has reshaped not only physics but the whole of human thinking. Its implications are so profound and many-sided that the history of science and philosophy may even be divided into two distinct parts: before and after the theory of relativity. Obviously the theory of relativity is not the only jump in the development of thinking. However it has meant the most important alteration of the physical and mathematical representation of Nature. It has brought about the interdependence of space and time and their dependence on the reference frame. It has demonstrated the dependence of mass, that is of inertia, on the frame, since it has appeared that mass is a function of velocity.

In this respect, Einstein's words are strikingly relevant: that the point in space where an event occurs has no physical meaning, neither does the moment when the event occurs; only the event itself is meaningful. Two events are not related by any absolute relation in space or any absolute relation in time; they are related by an absolute space-time relation (66). Highly accurate experiments have produced evidence that the relativistic model outclasses the prerelativistic one, although the latter is more accessible to common intuition.

The foundation of general relativity is special relativity, whose formalism is dominated by Minkowski's four-dimensional interpretation.

A.II.1. Special relativity

Formulated in 1905, the theory of special relativity has rewarded the efforts of numerous scientists (Lorentz, Poincaré, Einstein, Fitzgerald, etc.), to develop electrodynamics. Einstein's works gained outstanding authority because they were clear, daring and more independent of prerelativist physics.

At the basis of special relativity Einstein set two postulates:

(1) All phenomena occur identically in any inertial reference frame provided that the initial conditions are the same;

(2) The speed of light in empty space is a universal constant in all inertial frames, in all directions and does not depend either on the velocity of the source or on that of the observer. It is the greatest speed of propagation of information or influence from one point to another.

The first principle is a component of the axiomatics of classical mechanics. The second is specific to the theory of relativity and has been imposed through observation and experiment as an absolute necessity. This principle serves for Einstein to express the interdependence of space and time and their dependence on frame. Thus Einstein has taken a much greater step than one realizes at first sight towards the understanding of the laws of nature.

A.II.1.1. Synchronization of clocks. If one admits that in two frames that are not at relative rest the speed of light is expressed by the same constant, several consequences result, of which the first two are the following:

(a) time in one frame does <u>not</u> pass identically with time in another frame;

(b) in any inertial frame clocks can be and have to be synchronized.

To approach the problem of the synchronization of clocks it is useful to specify that:

(α) The space S of a reference frame Oxyz is the set of all points P at rest with respect to Oxyz. Obviously, S has to be a continuous set. The space S_1 of frame $O_1x_1y_1z_1$ is defined in the same way. The frames are considered Cartesian.

(β) In the space of an <u>inertial</u> frame no kinematic (inertial) or gravitational field of forces is observed. Consequently, if frames Oxyz and $O_1x_1y_1z_1$ are inertial, they are engaged in a rectilinear uniform translation with respect to each other.

The synchronization of clocks requires observance of the following conditions:

(1) All clocks attached to any point $P \in S$ run at the same speed. All of them will show the same time interval Δt between two events E_1 and E_2.

(2) The origins of time of those clocks attached to points $P \in S$ are so shifted with respect to each other that the observer at any point $P \in S$ records any event E at the same time t, no matter which clock is used: one at a point P or one attached to the origin O of the frame Oxyz.

Obviously the same conditions have to be fulfilled to synchronize clocks in space S_1 of frame $O_1x_1y_1z_1$.

If frames Oxyz and $O_1x_1y_1z_1$ are inertial, synchronization is possible. From now on they are considered inertial without specification; when noninertial frames are involved this will be stated.

The postulate regarding the invariance of the speed of light which is expressed by one <u>constant c</u> in frames Oxyz and $O_1x_1y_1z_1$ yields:

(a) If frames Oxyz and $O_1x_1y_1z_1$ are <u>not</u> at relative rest, the clocks in space S and S_1 run at different speeds.

(b) Two events E and E' which appear as simultaneous in S are no longer simultaneous in S_1, and conversely.

(c) Consider two events E_A and E_B recorded at t_A and t_B by the observer attached to S and at t_{1A} and t_{1B} by the observer attached to S_1; assume $t_B > t_A$. Then, because the speed of clocks at S and S_1 is different, it turns out that $t_B - t_A = \Delta t \neq t_{1B} - t_{1A} = \Delta t_1$. Moreover, although $\Delta t > 0$, it is possible that $\Delta t_1 > 0$, $\Delta t_1 = 0$ or $\Delta t_1 < 0$. That is, even the sequence of events may reverse on transfer from one frame to the other.

In conclusion, the group of relations which produce the transformation of coordinates for the transfer from one frame to the other has to produce the transformation of time as well, because time is not the same in S and S_1. In any frame, to locate an event, the space no longer suffices; a space-time location is needed. The transformation of space-time coordinates between two inertial frames is achieved by a set of relations referred to as the Lorentz-Einstein equations.

A.II.1.2. <u>The Lorentz-Einstein relations</u>. These relations were first advanced by Lorentz to ensure the invariance of the equations of electrodynamics and were then rigorously deduced by Einstein (the demonstrations are presumably known) as a consequence of the postulate regarding the constancy of the speed of light, c.

If one assumes that the axis O_1x_1 coincides with Ox and that both coincide with the direction of the relative translation of the frames, so that axis O_1Y_1 is parallel to Oy and O_1z_1 is parallel to Oz, these relations read:

$$x = \frac{x_1 - v\,t_1}{\sqrt{1 - \frac{v^2}{c^2}}};\ y = y_1;\ z = z_1;$$

$$t = \frac{t_1 - \frac{v}{c^2}\,x_1}{\sqrt{1 - \frac{v^2}{c^2}}}; \qquad \text{(A.1)}$$

and

$$x_1 = \frac{x + v\,t}{\sqrt{1 - \frac{v^2}{c^2}}};\ y_1 = y;\ z_1 = z;$$

$$t_1 = \frac{t + \frac{v}{c^2}\,x}{\sqrt{1 - \frac{v^2}{c^2}}}; \qquad \text{(A.2)}$$

where v is the velocity of Oxyz with respect to $O_1x_1y_1z_1$.

The particular choice of the relative motion of the frame does not restrict the generality of the treatment.

The eqns. (A.1) and (A.2), which relate the space-time coordinates x, y, z, t of event E in frame Oxyz to the coordinates x_1, y_1, z_1, t_1 of the same event E in frame $O_1x_1y_1z_1$, have important consequences. These affect the core of the problem rather than just its expression. From among their numerous consequences, those of outstanding importance are briefly considered next.

(a) Transformation of spatial interval. Consider a segment AB attached to frame Oxyz. An observer, also attached to Oxyz, measures the segment AB and finds its length ℓ.

Let now x_A, y_A, z_A be the coordinates of point A and x_B, y_B, z_B the coordinates of point B in the frame Oxyz. If the segment AB is parallel to Ox, the axis of relative translation of Oxyz and $O_1x_1y_1z_1$, the following relations hold:

$$x_B - x_A = \ell, \; y_B = y_A, \; z_B = z_A. \tag{A.3}$$

The problem now arises to find the length of AB found by an observer in frame $O_1x_1y_1z_1$.

Because this observer is not at rest with respect to the segment AB, a direct measurement of its length is not possible. The observer notes the event E_1, consisting of the presence of the end A of the segment at the point of coordinates x_{1A}, y_{1A}, z_{1A} at time t_{1A}, and the event E_{1B} which consists of the end B being at point of coordinates x_{1B}, y_{1B}, z_{1B} at time t_{1B} (in frame $O_1x_1y_1z_1$).

According to eqns. (A.1) and (A.3) one can write

$$x_B - x_A = \ell = \frac{x_{1B} - x_{1A} - v\,(t_{1B} - t_{1A})}{\sqrt{1 - \frac{v^2}{c^2}}}. \tag{A.4}$$

In order that $x_{1B} - x_{1A} = \ell_1$ be the length of the segment AB recorded by the observer in O_1, E_{1A} and E_{1B} have to be simultaneous, i.e., $t_{1A} = t_{1B}$. Then eqn. (A.4) becomes:

$$\ell = \frac{\ell_1}{\sqrt{1 - \frac{v^2}{c^2}}}; \quad \ell_1 = \ell \sqrt{1 - \frac{v^2}{c^2}}. \tag{A.5}$$

It turns out that the length of the segment AB is the greatest in its proper reference frame, and the length is then called the proper length. The proper frame of a body is the frame relative to which the body is at rest.

Likewise, if a segment MP lying on a straight line parallel to the translation direction in its proper frame has a proper length λ_1, then an observer bound to Oxyz records a length λ of the same segment:

$$\lambda = \lambda_1 \sqrt{1 - \frac{v^2}{c^2}} . \tag{A.6}$$

More generally, consider a segment CD along any direction whose proper frame is Oxyz. Let b_x, b_y, b_z be the projections of this segment on Ox, Oy, and Oz, respectively. The proper length of CD is

$$b = \sqrt{b_x^2 + b_y^2 + b_z^2}. \tag{A.7}$$

The length of CD noted by an observer bound to the frame $O_1x_1y_1z_1$ is:

$$b_1 = \sqrt{b_x^2 \left(1 - \frac{v^2}{c^2}\right) + b_y^2 + b_z^2}. \tag{A.8}$$

Consider now a line segment QL attached to $O_1x_1y_1z_1$ whose projections on O_1x_1, O_1y_1, O_1z_1 are δx_1, δy_1, δz_1, respectively; its proper length is

$$\delta_1 = \sqrt{\delta_{x_1}^2 + \delta_{y_1}^2 + \delta_{z_1}^2}, \tag{A.9}$$

and an observer bound to Oxyz records the following length of the same segment QL:

$$\delta = \sqrt{\delta_{x_1}^2 \left(1 - \frac{v^2}{c^2}\right) + \delta_{y_1}^2 + \delta_{z_1}^2}. \tag{A.10}$$

Eqns. (A.8) and (A.10) are the obvious consequences of eqns. (A.5) and (A.6) and of the fact that the projections of the segment perpendicular to the direction of translation do not change.

(b) The transformation of directions. The direction of a segment in the proper frame differs from that noted by an observer attached to another frame. Indeed, returning to segment CD, its direction cosines in the proper frame, Oxyz, are:

$$\cos\alpha = \frac{b_x}{\sqrt{b_x^2 + b_y^2 + b_z^2}},$$

$$\cos\beta = \frac{b_y}{\sqrt{b_x^2 + b_y^2 + b_z^2}}, \tag{A.11}$$

$$\cos\gamma = \frac{b_z}{\sqrt{b_x^2 + b_y^2 + b_z^2}}.$$

According to eqn. (A.8) the direction of the same segment CD recorded by the observer attached to $O_1x_1y_1z_1$ is expressed by the direction cosines:

$$\cos\alpha_1 = \frac{b_x\sqrt{1 - \frac{v^2}{c^2}}}{\sqrt{b_x^2\,(1 - \frac{v^2}{c^2}) + b_y^2 + b_z^2}},$$

$$\cos\beta_1 = \frac{b_y}{\sqrt{b_x\,(1 - \frac{v^2}{c^2}) + b_y^2 + b_z^2}}, \qquad \text{(A.12)}$$

$$\cos\gamma_1 = \frac{b_z}{\sqrt{b_x\,(1 - \frac{v^2}{c^2}) + b_y^2 + b_z^2}}.$$

Similarly, the segment QL considered above has the following direction cosines in its proper frame $O_1x_1y_1z_1$:

$$\cos\alpha_1' = \frac{\delta_{x_1}}{\sqrt{\delta_{x_1}^2 + \delta_{y_1}^2 + \delta_{z_1}^2}},$$

$$\cos\beta_1' = \frac{\delta_{y_1}}{\sqrt{\delta_{x_1}^2 + \delta_{y_1}^2 + \delta_{z_1}^2}}, \qquad \text{(A.13)}$$

$$\cos\gamma_1' = \frac{\delta_{z_1}}{\sqrt{\delta_{x_1}^2 + \delta_{y_1}^2 + \delta_{z_1}^2}}.$$

According to eqn. (A.10), for an observer attached to Oxyz, the direction of the same segment QL becomes:

$$\cos\alpha' = \frac{\delta_{x_1}\sqrt{1 - \frac{v^2}{c^2}}}{\sqrt{\delta_{x_1}^2 (1 - \frac{v^2}{c^2}) + \delta_{y_1}^2 + \delta_{z_1}^2}},$$

$$\cos\beta' = \frac{\delta_{y_1}}{\sqrt{\delta_{x_1}^2 (1 - \frac{v^2}{c^2}) + \delta_{y_1}^2 + \delta_{z_1}^2}}, \qquad (A.14)$$

$$\cos\gamma' = \frac{\delta_{z_1}}{\sqrt{\delta_{x_1}^2 (1 - \frac{v^2}{c^2}) + \delta_{y_1}^2 + \delta_{z_1}^2}}.$$

The direction of a straight line segment in its proper frame is the proper direction.

(c) The change of shape. The shape of a body is determined by the distances between pairs of its constitutive points and the directions determined by these points. Therefore, the relations that transform lengths and directions describe the change of shape on transfer from one frame Oxyz to any other, $O_1x_1y_1z_1$, or the reverse.

A body takes it proper shape in the proper frame; however this shape does not correspond to the shape noted by an observer who is not attached to the proper frame.

(d) The transformation of time intervals. Consider a point $P(x,y,z) \in S$, i.e., its coordinates x, y, z in Oxyz do not change. Consider further the event E_0 occurring at point P at time t_0 and the event E occurring at P at time $t > t_0$. Both times are recorded by an observer that belongs to space S (is attached to Oxyz).

The two moments t_0 and t define a time interval:

$$\Delta t = t - t_0. \tag{A.15}$$

The problem to be solved next is: what is the value of the time interval defined by the same two events, E_0 and E, recorded by another observer in space S_1 (attached to $O_1x_1y_1z_1$)?

According to eqns. (A.2) the moment t_0 in space S is recorded in space S_1 at time:

$$t_{01} = \frac{t_0 + \frac{v}{c^2} x}{\sqrt{1 - \frac{v^2}{c^2}}}, \tag{A.16}$$

and the moment t in space S corresponds to time:

$$t_1 = \frac{t + \frac{v}{c^2} x}{\sqrt{1 - \frac{v^2}{c^2}}} \tag{A.17}$$

in space S_1. This means that, for an observer in S_1, the time interval Δt_1 between events E_0 and E becomes:

$$\Delta t_1 = t_1 - t_{01} = \frac{t - t_0}{\sqrt{1 - \frac{v^2}{c^2}}} = \frac{\Delta t}{\sqrt{1 - \frac{v^2}{c^2}}}. \qquad (A.18)$$

It is assumed now that at point $Q(x_1, y_1, z_1) \in S_1$ the events E_{01} and E_1 occur, recorded at time t_{01} and $t'_1 > t'_{01}$ by an observer in space S_1. Another observer located in S notes the same events E_{01} and E_1 at time t'_0 and t'. The intervals between these events are $\Delta t'_1 = t'_1 - t'_{01}$ in S_1 and $\Delta t' = t' - t'_0$ in S.

By a similar argument to the above, eqns. (A.1) imply that:

$$\Delta t' = \frac{\Delta t'_1}{\sqrt{1 - \frac{v^2}{c^2}}}. \qquad (A.19)$$

The proper time interval of a reference frame is the duration measured by an observer bound to the frame between two events which occur at the same point also attached to the frame in question.

Eqns. (A.18) and (A.19) highlight the conclusion that if n observers, each attached to a reference frame, measure the time interval between two events, the proper time interval is the shortest one. Moreover, the higher the velocity of the relative motion of the observer's frame with respect to the frame in which the time interval is proper, the larger the value of the time interval as recorded by the observer.

(e) The definition of 4-vectors. To reach a concise, adequate formulation, the space-time coordinates of an event E in frame Oxyz may be denoted by:

$$x = w_1;\ y = w_2;\ z = w_3;\ ict = w_4. \qquad (A.20)$$

Similarly, the coordinates of the same event E in frame $O_1x_1y_1z_1$ are:

$$x_1 = u_1;\ y_1 = u_2;\ z_1 = u_3;\ ict_1 = u_4. \qquad (A.21)$$

In eqns. (A.20) and (A.21), c is the speed of light and $i = \sqrt{-1}$ is the imaginary unit.

Then the Lorentz-Einstein relations become:

$$w_k = \sum_{s=1}^{4} \alpha_{sk} u_s;\ k = 1,2,3,4; \qquad (A.22)$$

and

$$u_k = \sum_{s=1}^{4} \nu_{sk} w_s;\ k = 1,2,3,4. \qquad (A.23)$$

The coefficients α_{sk} and ν_{sk} make up the matrices:

$$\|\alpha_{sk}\| = \left\| \begin{matrix} \frac{1}{\sqrt{1-\beta^2}} & 0 & 0 & \frac{i\beta}{\sqrt{1-\beta^2}} \\ 0 & 1 & 0 & 0 \\ 0 & 0 & 1 & 0 \\ \frac{-i\beta}{\sqrt{1-\beta^2}} & 0 & 0 & \frac{1}{\sqrt{1-\beta^2}} \end{matrix} \right\| \qquad (A.24)$$

and

$$\left\| \nu_{sk} \right\| = \left\| \begin{matrix} \frac{1}{\sqrt{1 - \beta^2}} & 0 & 0 & \frac{-i\beta}{\sqrt{1 - \beta^2}} \\ 0 & 1 & 0 & 0 \\ 0 & 0 & 1 & 0 \\ \frac{i\beta}{\sqrt{1 - \beta^2}} & 0 & 0 & \frac{1}{\sqrt{1 - \beta^2}} \end{matrix} \right\|, \quad (A.25)$$

where $\beta = \frac{v}{c}$.

With that choice of the relative position of axes of frames that leads to the expression (A.1), (A.2) of the Lorentz-Einstein relations, any physical quantity, expressed by four components which, on transfer from one frame to another, transform according to eqns. (A.22) and (A.23), where the coefficients are contained in matrices (A.24) and (A.25), is referred to as a 4-vector.

Thus, a 4-vector is a vector in a four-dimensional space; its components transform according to Lorentz-Einstein relations.

Obviously, any other choice of axes of the inertial reference frames would lead to expressions for the matrices which differ from (A.24) and (A.25). However, if we apply consistently to the matrices that relate reference frames the constraints stemming from Einstein's postulates, this gives the set of these matrices a privileged mathematical quality: as is well known, they make up a group - the Lorentz group, dealt with through an extensive literature.

The fact that in this book we do not take advantage of the general mathematical facilities provided by the Lorentz groups as such is not a matter of the economy of thinking only. Indeed, one cannot stress enough the importance of the fact that the results of special relativity do not depend on the concrete choice of frames, but only on the inertial character of these, subject to Einstein's postulates. Highlighting this is much in keeping with the very spirit of relativity. In fact, the consequences of such an attitude will already become apparent in the next paragraph.

A.II.1.3. The invariant of Lorentz-Einstein transformations. Any group of transformation relations has to observe the invariance of some quantities which are called the invariants of the transformation. Because they do not change when the frame changes, they suggest the possibility of giving the laws of nature unitary and general expressions, regardless of the frame employed.

The invariants depend on the group of relations which produce the transformation of coordinates on passing from one frame to the other.

Consider two events E and E' occurring at two points M and M' at two different times; let x, y, z, t and x_1, y_1, z_1, t_1 be the space-time coordinates of E in Oxyz and $O_1x_1y_1z_1$, respectively, and x', y', z', t' and x_1', y_1', z_1', t_1' the space-time coordinates of E' in Oxyz and $O_1x_1y_1z_1$, respectively.

The following notations are adopted:

$$\begin{aligned} &x' - x = \Delta x; \; y' - y = \Delta y; \; z' - z = \Delta z; \\ &t' - t = \Delta t; \\ &x_1' - x_1 = \Delta x_1; \; y_1' - y_1 = \Delta y_1; \\ &z_1' - z_1 = \Delta z_1; \; t_1' - t_1 = \Delta t_1. \end{aligned} \quad (A.26)$$

Using (A.24) and (A.25), it is easy to check that, whatever the space-time coordinates of the two events in frames Oxyz and $O_1x_1y_1z_1$, the following equation holds:

$$c^2 \Delta t^2 - (\Delta x^2 + \Delta y^2 + \Delta z^2) = c^2 \Delta t_1^2 - (\Delta x_1^2 + \Delta y_1^2 + \Delta z_1^2). \tag{A.27}$$

Thus an invariant of the Lorentz-Einstein transformation is:

$$\Delta s = [c^2 \Delta t^2 - (\Delta x^2 + \Delta y^2 + \Delta z^2)]^{\frac{1}{2}} = [c^2 \Delta t_1^2 - (\Delta x_1^2 + \Delta y_1^2 + \Delta z_1^2)]^{\frac{1}{2}}, \tag{A.28}$$

and this is referred to as the <u>space-time interval</u> between events E and E'.

It is worth noting that the distance between points M and M', which is an invariant of the Galilei transformation of coordinates, ceases to be an invariant of the Lorentz-Einstein transformation.

Obviously, in the case of space-time infinitesimal intervals, one can write:

$$ds = [c^2 dt^2 - (dx^2 + dy^2 + dz^2)]^{\frac{1}{2}} = [c^2 dt_1^2 - (dx_1^2 + dy_1^2 + dz_1^2)]^{\frac{1}{2}}. \tag{A.29}$$

The space-time intervals between two events requires further analysis, which is given next.

A.II.1.4. <u>Classification of space-time intervals</u>. Consider a frame Oxyz, out of the set of inertial frames, in whose space events E and E' occur at the same point

(M = M'). According to (A.26) this implies the equations $\Delta x = \Delta y = \Delta z = 0$ which, according to eqn. (A.28), means:

$$\Delta s^2 = c^2 \Delta t^2 > 0. \tag{A.30}$$

Such a space-time interval is called a time-like interval.

The invariance of Δs leads to the conclusion that in any other inertial frame $O_1 x_1 y_1 z_1$ the following inequality holds:

$$\Delta s^2 = c^2 \Delta t_1^2 - (\Delta x_1^2 + \Delta y_1^2 + \Delta z_1^2) > 0. \tag{A.31}$$

However, because $c^2 \Delta t_1^2 > \Delta x_1^2 + \Delta y_1^2 + \Delta z_1^2$, it turns out that points M and M' where events E and E' occur ($|MM'| = (\Delta x_1^2 + \Delta y_1^2 + \Delta z_1^2)^{\frac{1}{2}}$) lie inside a sphere of radius $R_1 = c\, \Delta t_1$ whose centre is at one of the points ('e.g.' M).

Now, since it was admitted that c is the maximum speed that the propagation of any influence (interaction) may assume, it follows that causality relationships may exist between the points inside the sphere. Thus, the event E may be the cause of event E' (but not necessarily is) when the two events are separated by a time-like space-time interval.

If there is a frame, e.g. Oxyz, wherein E and E' appear as simultaneous, then notations (A.26) yield $\Delta t = 0$; and from eqn. (A.28) it results that

$$\Delta s^2 = -(\Delta x^2 + \Delta y^2 + \Delta z^2) < 0. \tag{A.32}$$

This means that the observer in another frame,

e.g., $O_1x_1y_1z_1$ notes that

$$\Delta s^2 = c^2 \Delta t_1^2 - (\Delta x_1^2 + \Delta y_1^2 + \Delta z_1^2) < 0. \qquad (A.33)$$

This is a space-like space-time interval.

Thus $\Delta x_1^2 + \Delta y_1^2 + \Delta z_1^2 > c^2 \Delta t_1^2$, which means that the points M and M' where events E and E' occur lie so far apart that one or other is outside the sphere mentioned above; consequently, if events E and E' are separated by a space-like interval, any causality relationship between them is ruled out.

Finally, assume there is a frame (perhaps Oxyz) in which events E and E' appear as simultaneous ($t = t'$) and at points that coincide ($M = M'$); then eqn. (A.28) shows that $\Delta s = 0$.

As a consequence, in another frame $O_1x_1y_1z_1$ the following equation holds:

$$\Delta s^2 = c \Delta t_1^2 - (\Delta x_1^2 + \Delta y_1^2 + \Delta z_1^2) = 0. \qquad (A.34)$$

The events separated by such zero space-time intervals are simultaneous with the arrival at the points where they occur of the same light wave (or any other influence, perturbation or interaction which propagates with the speed of light).

Note. Taking into account that frames Oxyz and $O_1x_1y_1z_1$ have parallel homologous axes and O_1x_1 coincides with Ox and with their direction of relative translation, it follows that $\Delta y = \Delta y_1$; $\Delta z = \Delta z_1$ and then eqn. (A.28) yields:

$$c^2 \Delta t^2 - \Delta x^2 = c^2 \Delta t_1^2 - \Delta x_1^2. \qquad (A.35)$$

In case of time-like intervals $\Delta x = 0$ and the transformations (A.2) give:

$$\Delta t_1 = \frac{\Delta t}{\sqrt{1 - \frac{v^2}{c^2}}} . \tag{A.36}$$

This means that $\Delta t > 0$ implies $\Delta t_1 > 0$, and conversely. Conclusion: in the case of time-like intervals, the succession of events is the same in all frames.

If the intervals are space-like, $\Delta t = 0$ and the same eqns. (A.2) yield:

$$\Delta t_1 = \frac{\frac{v}{c^2} \Delta x}{\sqrt{1 - \frac{v^2}{c^2}}} . \tag{A.37}$$

It is equally possible that $\Delta x > 0$ or $\Delta x < 0$, so that it is possible that $\Delta t_1 > 0$ and $\Delta t_1 < 0$. Thus in frame $O_1x_1y_1z_1$ any succession of events E and E' is possible.

A.II.1.5. Minkowski space. The notations introduced by eqns. (A.20) and (A.21) allow one to conceive a fictitions four-dimensional space (the fourth coordinate being time). Such a space is referred to as Minkowski space, to honour the scientist who introduced it in the calculations and reasoning of the theory of relativity.

According to eqns. (A.20) and (A.21) the relations (A.28) and (A.29) may be rewritten as:

$$\Delta s^2 = - \sum_{k=1}^{4} \Delta w_k^2 = - \sum_{k=1}^{4} \Delta u_k^2 \tag{A.38}$$

and

$$ds^2 = - \sum_{k=1}^{4} dw_k^2 = - \sum_{i=1}^{4} du_i^2. \qquad (A.39)$$

Any event occurring at a given time in three-dimensional physical space corresponds to one point in Minkowski space. A continuous succession of events in the physical space corresponds in Minkowski space to a curve called a world line.

The Minkowski space facilitates expression and therefore allows some new developments.

However, it is worth noting that the interpretation of time as a fourth dimension of space, while mathematically consistent and extremely useful, still remains physically inadequate and frustrating.

Time may not be taken as a dimension of space because:

(a) the time coordinate in Minkowski space contains the imaginary unit $i = \sqrt{-1}$;

(b) the main feature of time is its irreversibility, which makes it unlike any spatial dimension.

Thus time remains a basic distinct quantity, and Minkowski space is purely fictitious.

Consider now two events E_0 (x_0, y_0, z_0, t_0) and E (x, y, z, t) whose space-time coordinates are given in the reference frame Oxyz employed above.

Because

$$\Delta s^2 = c^2 (t - t_0)^2 - [(x - x_0)^2 + (y - y_0)^2 + (z - z_0)^2],$$

the notations (A.20) allow one to write:

$$\Delta s^2 = -[(w_1 - w_{0_1})^2 + (w_2 - w_{0_2})^2 + (w_3 - w_{0_3})^2 + (w_4 - w_{0_4})^2], \tag{A.40}$$

with $w_{0_1} = x_0$; $w_{0_2} = y_0$; $w_{0_3} = z_0$; $w_{0_4} = ict_0$.

This means that $\Delta s = 0$ represents in Minkowski space a conical surface called the light cone. The vertex of this conical surface in Minkowski space is the point A $(w_{0_1}, w_{0_2}, w_{0_3}, w_{0_4})$ which corresponds to event E (x_0, y_0, z_0, t_0) in the physical space S of frame Oxyz.

It is readily seen that all points A (w_1, w_2, w_3, w_4) inside the light cone possess the following feature: substitution of their coordinates in eqn. (A.40) yields $\Delta s^2 > 0$. Thus Δs is a time-like interval and events E (x, y, z, t) in the physical space S of frame Oxyz corresponding to points A (w_1, w_2, w_3, w_4) are: (a) either ulterior to event E_0, if $t > t_0$, and their set makes up the future in respect of E_0; (b) or anterior to event E, if $t < t_0$, and their set forms the past in respect of E_0. This is so whatever the inertial frame Oxyz.

The points A (w_1, w_2, w_3, w_4) outside the light cone in Minkowski space require that $\Delta s^2 < 0$, so that they correspond to events E (x, y, z, t) which together with event E_0 (x_0, y_0, z_0, t_0) define space-like intervals. Therefore, whatever E is chosen, in some Oxyz frames it is anterior to event E_0 and in other frames it is ulterior to E_0.

Note. Instead of light cone, influence cone or interaction cone may be more adequate. These names would emphasize better the delimitation of events that may have causality relationships from those that may not have. This delimitation is achieved by the conical surface:

$$(w_1 - w_{0_1})^2 + (w_2 - w_{0_2})^2 +$$

$$(w_3 - w_{0_3})^2 + (w_4 - w_{0_4})^2 = 0.$$

A.II.1.6. Transformation of velocity components. Consider the material point P and the two frames Oxyz and $O_1x_1y_1z_1$ employed before. P generally moves with respect to both frames; but in particular cases it may be at rest in one of the frames.

Let $\bar{i}$, $\bar{j}$, $\bar{k}$ be the unit vectors along Ox, Oy, Oz and $\bar{i}_1$, $\bar{j}_1$, $\bar{k}_1$ along O_1x_1, O_1y_1, O_1z_1; let $\bar{\Omega}$ be the velocity of P with respect to Oxyz and $\bar{\Omega}_1$ the velocity of P with respect to $O_1x_1y_1z_1$.

Then, the following equations hold:

$$\bar{\Omega} = \bar{i}\,\Omega_x + \bar{j}\,\Omega_y + \bar{k}\,\Omega_z, \tag{A.41}$$

$$\Omega_1 = \bar{i}_1\,\Omega x_1 + \bar{j}_1\,\Omega y_1 + \bar{k}_1\,\Omega z_1, \tag{A.42}$$

where:

$$\Omega_x = \frac{dx}{dt};\ \Omega_y = \frac{dy}{dt};\ \Omega_z = \frac{dz}{dt} \tag{A.43}$$

and:

$$\Omega_{x_1} = \frac{dx_1}{dt_1};\ \Omega_{y_1} = \frac{dy_1}{dt_1};\ \Omega_{z_1} = \frac{dz_1}{dt_1}. \tag{A.44}$$

The problem now arises, how to find the relation between the velocity of point P in Oxyz and $O_1x_1y_1z_1$.

The Lorentz-Einstein transformations yield:

$$dx = \frac{dx_1 - v\,dt_1}{\sqrt{1 - \frac{v^2}{c^2}}};\; dy = dy_1;\; dz = dz_1;$$

$$dt = \frac{dt_1 - \frac{v}{c^2}\,dx_1}{\sqrt{1 - \frac{v^2}{c^2}}} \tag{A.46}$$

and

$$dx_1 = \frac{dx + v\,dt}{\sqrt{1 - \frac{v^2}{c^2}}};\; dy_1 = dy;\; dz_1 = dz;$$

$$dt_1 = \frac{dt + \frac{v}{c^2}\,dx}{\sqrt{1 - \frac{v^2}{c^2}}}. \tag{A.46}$$

Hence, by straightforward calculation, one gets:

$$\Omega_x = \frac{dx}{dt} = \frac{\Omega_{x_1} - v}{1 - \frac{v}{c^2}\,\Omega_{x_1}},$$

$$\Omega_y = \frac{dy}{dt} = \frac{\Omega_{y_1}\sqrt{1 - \frac{v^2}{c^2}}}{1 - \frac{v}{c^2}\,\Omega_{x_1}}, \tag{A.47}$$

$$\Omega_z = \frac{dz}{dt} = \frac{\Omega_{z_1}\sqrt{1 - \frac{v^2}{c^2}}}{1 - \frac{v}{c^2}\,\Omega_{x_1}}$$

as well as

$$\Omega_{x_1} = \frac{dx_1}{dt_1} = \frac{\Omega_x + v}{1 + \frac{v}{c^2}\,\Omega_x},$$

$$\Omega_{y_1} = \frac{dy_1}{dt_1} = \frac{\Omega_y\sqrt{1 - \frac{v^2}{c^2}}}{1 + \frac{v}{c^2}\,\Omega_x}, \tag{A.48}$$

$$\Omega_{z_1} = \frac{dz_1}{dt_1} = \frac{\Omega_z\sqrt{1 - \frac{v^2}{c^2}}}{1 + \frac{v}{c^2}\,\Omega_x}.$$

Considering now the magnitude of the vectors $\bar{\Omega}$ and $\bar{\Omega}_1$, they are:

$$|\bar{\Omega}| = \sqrt{\Omega_x^2 + \Omega_y^2 + \Omega_z^2}, \tag{A.49}$$

$$|\bar{\Omega}_1| = \sqrt{\Omega_{x_1}^2 + \Omega_{y_1}^2 + \Omega_{z_1}^2}, \tag{A.50}$$

and combination of (A.47) with (A.49) and (A.48) with (A.50) gives

$$|\bar{\Omega}| = \frac{\sqrt{(\Omega_1^2 - \Omega_{x_1}^2)(1 - \frac{v^2}{c^2}) + v^2(1 - \frac{\Omega_{x_1}}{v})^2}}{1 - \frac{v}{c^2}\Omega_{x_1}}, \quad (A.51)$$

$$|\bar{\Omega}_1| = \frac{\sqrt{(\Omega^2 - \Omega_x^2)(1 - \frac{v^2}{c^2}) + v^2(1 + \frac{\Omega}{v})^2}}{1 + \frac{v}{c^2}\Omega_x}. \quad (A.52)$$

The direction cosines of vectors $\bar{\Omega}$ and $\bar{\Omega}_1$ are (obviously, each in its frame):

$$\cos\alpha = \frac{\Omega_x}{|\bar{\Omega}|}; \; \cos\beta = \frac{\Omega_y}{|\bar{\Omega}|}; \; \cos\gamma = \frac{\Omega_z}{|\bar{\Omega}|} \quad (A.53)$$

and

$$\cos\alpha_1 = \frac{\Omega_{x_1}}{|\bar{\Omega}_1|}; \; \cos\beta_1 = \frac{\Omega_{y_1}}{|\bar{\Omega}_1|}; \; \cos\gamma_1 = \frac{\Omega_{z_1}}{|\bar{\Omega}_1|}. \quad (A.54)$$

Eqns. (A.47), (A.51) and (A.53) yield:

$$\cos\alpha = \frac{\cos\alpha_1 - \frac{v}{\Omega_1}}{\sqrt{(1-\cos^2\alpha_1)(1 - \frac{v^2}{c^2}) + (\frac{v}{\Omega_1} - \cos\alpha_1)^2}},$$

$$\cos\beta = \frac{\sqrt{1 - \frac{v^2}{c^2}}\cos\beta_1}{\sqrt{(1-\cos^2\alpha_1)(1 - \frac{v^2}{c^2}) + (\frac{v}{\Omega_1} - \cos\alpha_1)^2}}, \quad (A.55)$$

$$\cos\gamma = \frac{\sqrt{1-\frac{v^2}{c^2}}\cos\gamma_1}{\sqrt{(1-\cos^2\alpha_1)(1-\frac{v^2}{c^2}) + (\frac{v}{\Omega_1}-\cos\alpha_1)^2}}.$$

Now eqns. (A.48), (A.52) and (A.54) together suggest the conclusion:

$$\cos\alpha_1 = \frac{\cos\alpha + \frac{v}{\Omega}}{\sqrt{(1-\cos^2\alpha)(1-\frac{v^2}{c^2}) + (\frac{v}{\Omega}+\cos\alpha)^2}},$$

$$\cos\beta_1 = \frac{\sqrt{1-\frac{v^2}{c^2}}\cos\beta}{\sqrt{(1-\cos^2\alpha)(1-\frac{v^2}{c^2}) + (\frac{v}{\Omega}+\cos\alpha)^2}}, \quad \text{(A.56)}$$

$$\cos\gamma_1 = \frac{\sqrt{1-\frac{v^2}{c^2}}\cos\gamma}{\sqrt{(1-\cos^2\alpha)(1-\frac{v^2}{c^2}) + (\frac{v}{\Omega}+\cos\alpha)^2}},$$

with $\bar{\Omega}$ and $\bar{\Omega}_1$ the velocities of P in Oxyz and $O_1x_1y_1z_1$, respectively.

Note. If $\Omega_x = c$ and $\Omega_y = \Omega_z = 0$, eqns. (A.48) yield:

$$\Omega_{x_1} = \frac{c+v}{1+\frac{v}{c^2}c} = c; \quad \Omega_{y_1} = 0; \quad \Omega_{z_1} = 0; \quad \text{(A.57)}$$

and, if $\Omega_{x_1} = c$ and $\Omega_{y_1} = \Omega_{z_1} = 0$, eqns. (A.47) readily give:

$$\Omega_x = \frac{c - v}{1 - \frac{v}{c^2} c} = c; \; \Omega_y = 0; \; \Omega_z = 0. \tag{A.58}$$

More generally, from eqns. (A.51) and (A.52) for $|\bar{\Omega}_1| = c$, it follows, after elementary calculations, that $|\bar{\Omega}| = c$, and conversely $|\bar{\Omega}| = c$ yields $|\bar{\Omega}_1| = c$. This is a direct consequence of the postulate regarding the uniquely constant character of the speed of light.

Likewise, substitution of $\Omega_1 = c$ in eqns. (A.55) and $\Omega = c$ in eqns. (A.56) give:

$$\cos\alpha = \frac{c \cos\alpha_1 - v}{c - v \cos\alpha_1},$$

$$\cos\beta = \frac{\sqrt{1 - \frac{v^2}{c^2}} \cos\beta_1}{c - v \cos\alpha_1}, \tag{A.59}$$

$$\cos\gamma = \frac{\sqrt{1 - \frac{v^2}{c^2}} \cos\gamma_1}{c - v \cos\alpha_1}$$

and

$$\cos\alpha_1 = \frac{c \cos\alpha + v}{c + v \cos\alpha},$$

$$\cos\beta_1 = \frac{\sqrt{1 - \frac{v^2}{c^2}} \cos\beta}{c + v \cos\alpha}, \tag{A.60}$$

$$\cos\gamma_1 = \frac{\sqrt{1 - \frac{v^2}{c^2}} \cos\gamma}{c + v \cos\alpha},$$

respectively.

This is the relation between the direction of a light ray in Oxyz and the direction of the same ray in $O_1x_1y_1z_1$; directions are given by direction cosines.

It turns out that a body moves at a velocity $\bar{\Omega}$ with respect to Oxyz and $\bar{\Omega}_1$ with respect to $O_1x_1y_1z_1$; however if $|\bar{\Omega}| = c$ then $|\bar{\Omega}_1| = c$, and the direction of vector $\bar{\Omega}$ in Oxyz differs from that of vector $\bar{\Omega}_1$ in the space of frame $O_1x_1y_1z_1$.

A.II.1.7. The velocity and acceleration 4-vectors. The way the proper time of a reference system was defined suggests that one can define the proper time of a body. The proper time of a body is the proper time of the frame attached to the body, i.e., the time shown by a clock bound to the body.

Whatever the motion of material point P with respect to a frame Oxyz, elementary (infinitesimal) portions of it may be considered uniform rectilinear motions.

Consider Ox'y'z', a reference frame attached to point P. Generally, because P's motion with respect to Oxyz can be any motion, frames O'x'y'z' and Oxyz are not in inertial relative motion. However, the assimilation of elementary segments of P's motion to a uniform rectilinear motion implies the assimilation, on the same segments, of the relative motion of O'x'y'z' and Oxyz with an inertial motion.

Consider now two events E and E' consisting of the transit of the moving body P through two infinitely close positions: $P_0(x,y,z) \in S$ and $P_0'(x + dx,\ y + dy,\ z + dz) \in S$ where S stands for the space of frame Oxyz.

Events E and E' define a time interval which, measured by a clock attached to P, is $d\sigma$, and, by one attached to S, is dt.

Obviously $d\sigma$ is the proper time of the material point P, i.e., of the frame O'x'y'z'.

Because the intervals in question are infinitely small, the motion of O'x'y'z' may be taken as inertial. Thus, eqns. (A.18) and (A.19) yield:

$$dt = \frac{d\sigma}{\sqrt{1 - \frac{\Omega^2}{c^2}}}, \qquad (A.61)$$

where Ω is the magnitude of the velocity of P relative to xyz.

The space-time interval (with invariant character) delimited by events E and E' is:

$$ds = \sqrt{c^2dt^2 - (dx^2 + dy^2 + dz^2)}. \qquad (A.62)$$

Now we use the notations (A.20) and define the position 4-vector of event E in the frame Oxyz as a function $\bar{\rho}$ (w_1, w_2, w_3, w_4) of the space-time coordinates w_1, w_2, w_3, w_4 of event E in Oxyz.

Consider the reference frame $O_1x_1y_1z_1$ (which was already utilized above): according to eqns. (A.21), the position 4-vector of the event E in $O_1x_1y_1z_1$ is $\bar{\rho}_1(u_1, u_2, u_3, u_4)$.

The two events E and E', consisting of P's transit through positions $P_0(x,y,z) \in S$ and $P_0'(x + dx, y + dy, z + dz) \in S$, define the variation:

$$d\rho^2 = dw_1^2 + dw_2^2 + dw_3^2 + dw_4^2 =$$
$$dx^2 + dy^2 + dz^2 - c^2dt^2 = - ds^2 \qquad (A.63)$$

which is the square of the elementary 4-vector $d\bar{\rho}$.

Taking into account eqns. (A.22) and (A.23), it follows that:

$$dw_k = \sum_{s=1}^{4} \alpha_{sk}\, du_s; \quad k = 1,2,3,4 \tag{A.64}$$

and

$$du_k = \sum_{s=1}^{4} \nu_{sk}\, dw_s; \quad k = 1,2,3,4 \tag{A.65}$$

where coefficients α_{sk} and ν_{sk} are given by the matrices (A.24) and (A.25). The eqns. (A.64) and (A.65) relate the components of the elementary 4-vectors $d\bar{\rho}(dw_1, dw_2, dw_3, dw_4)$ and $d\bar{\rho}_1(du_1, du_2, du_3, du_4)$.

Considering the elementary 4-vector $d\bar{\rho}_1$, it is obvious that:

$$\begin{aligned} d\rho_1^2 &= du_1^2 + du_2^2 + du_3^2 + du_4^2 = \\ &dx_1^2 + dy_1^2 + dz_1^2 - c^2\, dt_1^2 = -\, ds^2, \end{aligned} \tag{A.66}$$

with x_1, y_1, z_1, t_1 and $x_1 + dx_1$, $y_1 + dy_1$, $z_1 + dz_1$, $t_1 + dt_1$ standing for the space-time coordinates of events E and E' in $O_1x_1y_1z_1$. Eqns. (A.63) and (A.64), which are equivalent to $d\rho^2 = d\rho_1^2$, reflect the invariance of the space-time interval ds.

The velocity 4-vector is defined as the derivative of the position 4-vector with respect to the proper time. That is:

$$\bar{q} = \frac{d\bar{\rho}}{d\sigma} \tag{A.67}$$

is the velocity 4-vector of the motion of the body P with respect to the frame Oxyz; its components are:

$$q_1 = \frac{dw_1}{d\sigma} = \frac{dx}{d\sigma}; \; q_2 = \frac{dw_2}{d\sigma} = \frac{dy}{d\sigma};$$
$$q_3 = \frac{dw_3}{d\sigma} = \frac{dz}{d\sigma}; \; q_4 = \frac{dw_4}{d\sigma} = ic\,\frac{dt}{d\sigma}. \tag{A.68}$$

Likewise:

$$\bar{Q} = \frac{d\bar{\rho}_1}{d\sigma} \tag{A.69}$$

is the velocity 4-vector of the motion of the same body P respect to $O_1x_1y_1z_1$. Its components are:

$$Q_1 = \frac{du_1}{d\sigma} = \frac{dx_1}{d\sigma}; \; Q_2 = \frac{du_2}{d\sigma} = \frac{dy_1}{d\sigma};$$

$$Q_3 = \frac{du_3}{d\sigma} = \frac{dz_1}{d\sigma}; \; Q_4 = \frac{du_4}{d\sigma} = ic\,\frac{dt_1}{d\sigma}. \tag{A.70}$$

Eqns. (A.64) and (A.65) yield immediately:

$$q_k = \sum_{s=1}^{4} \alpha_{sk}\,Q_s,$$
$$k = 1,2,3,4, \tag{A.71}$$
$$Q_k = \sum_{s=1}^{4} \nu_{sk}\,q_s,$$

which display the 4-vector character of $\bar{q}$ and $\bar{Q}$.

Now from eqn. (A.61) it follows that:

$$q_1 = \frac{dw_1}{d\sigma} = \frac{dx}{d\sigma} = \frac{dx}{dt} \cdot \frac{dt}{d\sigma} = \frac{\Omega_x}{\sqrt{1 - \frac{\Omega^2}{c^2}}},$$

$$q_2 = \frac{dw_2}{d\sigma} = \frac{dy}{d\sigma} = \frac{dy}{dt} \cdot \frac{dt}{d\sigma} = \frac{\Omega_y}{\sqrt{1 - \frac{\Omega^2}{c^2}}}, \qquad (A.72)$$

$$q_3 = \frac{dw_3}{d\sigma} = \frac{dz}{d\sigma} = \frac{dz}{dt} \cdot \frac{dt}{d\sigma} = \frac{\Omega_z}{\sqrt{1 - \frac{\Omega^2}{c^2}}},$$

$$q_4 = \frac{dw_4}{d\sigma} = \frac{d}{d\sigma}(ict) = \frac{ic}{\sqrt{1 - \frac{\Omega^2}{c^2}}}$$

where $\bar{\Omega}$, with components Ω_x, Ω_y, Ω_z, is the velocity of point P in Oxyz.

If Ω_1, with components Ω_{x_1}, Ω_{y_1}, Ω_{z_1}, is the velocity of P in $O_1x_1y_1z_1$, then, on the one hand:

$$dt_1 = \frac{d\sigma}{\sqrt{1 - \frac{\Omega_1^2}{c^2}}} \qquad (A.73)$$

and, on the other hand:

$$Q_1 = \frac{du_1}{d\sigma} = \frac{dx_1}{d\sigma} = \frac{dx_1}{dt_1} \cdot \frac{dt_1}{d\sigma} = \frac{\Omega_{x_1}}{\sqrt{1 - \frac{\Omega_1^2}{c^2}}}, \qquad (A.74)$$

$$Q_2 = \frac{du_2}{d\sigma} = \frac{dy_1}{d\sigma} = \frac{dy_1}{dt_1} \cdot \frac{dt_1}{d\sigma} = \frac{\Omega_{y_1}}{\sqrt{1 - \frac{\Omega_1^2}{c^2}}},$$

$$Q_3 = \frac{du_3}{d\sigma} = \frac{dz_1}{d\sigma} = \frac{dz_1}{dt_1} \cdot \frac{dt_1}{d\sigma} = \frac{\Omega_{z_1}}{\sqrt{1 - \frac{\Omega_1^2}{c^2}}},$$

$$Q_4 = \frac{du_4}{d\sigma} = \frac{d}{d\sigma}(ict_1) = \frac{ic}{\sqrt{1 - \frac{\Omega_1^2}{c^2}}}.$$

Eqns. (A.72) and (A.74) allow one to find out that:

$$q^2 = \left(\frac{dw_1}{d\sigma}\right)^2 + \left(\frac{dw_2}{d\sigma}\right)^2 + \left(\frac{dw_3}{d\sigma}\right)^2 + \left(\frac{dw_4}{d\sigma}\right)^2 = -c^2 \tag{A.75}$$

and

$$Q^2 = \left(\frac{du_1}{d\sigma}\right)^2 + \left(\frac{du_2}{d\sigma}\right)^2 + \left(\frac{du_3}{d\sigma}\right)^2 + \left(\frac{du_4}{d\sigma}\right)^2 = -c^2, \tag{A.76}$$

which evidences the invariance of the magnitude of the velocity 4-vector.

It is interesting to note that, in particular, for $P \in S$, $\bar{\Omega} = 0$ and, consequently,

$$\frac{dw_1}{d\sigma} = \frac{dw_2}{d\sigma} = \frac{dw_3}{d\sigma} = 0; \quad \frac{dw_4}{d\sigma} = ic. \tag{A.77}$$

These equations are interpreted from a physical standpoint as follows: there always exists a variation of the time coordinate of a body, even in the frame in which its position does not change. In other words, time never stops.

Very often the following notations are utilized:

$$\bar{\rho} = \bar{\rho}\,(\bar{r},\ ict),$$
$$\bar{\rho}_1 = \bar{\rho}_1\,(\bar{r}_1,\ ict), \tag{A.78}$$

where $\bar{r}$ and $\bar{r}_1$ stand for the position vectors of point P in Oxyz and $O_1x_1y_1z_1$, respectively, as well as:

$$\bar{q} = \bar{q}\left(\frac{\bar{\Omega}}{\sqrt{1 - \frac{\Omega^2}{c^2}}},\ \frac{i\,c}{\sqrt{1 - \frac{\Omega^2}{c^2}}}\right),$$
$$\bar{Q} = \bar{Q}\left(\frac{\bar{\Omega}_1}{\sqrt{1 - \frac{\Omega_1^2}{c^2}}},\ \frac{i\,c}{\sqrt{1 - \frac{\Omega_1^2}{c^2}}}\right). \tag{A.79}$$

The acceleration 4-vector $\bar{\varphi}$ of the moving body P in the space S of a frame Oxyz is defined by the equation:

$$\bar{\varphi} = \frac{d\bar{q}}{d\sigma} = \frac{d^2\bar{\rho}}{d\sigma^2} \tag{A.80}$$

The acceleration 4-vector $\bar{\phi}$ of the same body P in the space S_1 of a frame $O_1x_1y_1z_1$ is defined by the equation:

$$\bar{\phi} = \frac{d\bar{Q}}{d\sigma} = \frac{d^2\bar{\rho}_1}{d\sigma^2}. \tag{A.81}$$

Thus the components of the 4-vector $\bar{\varphi}$ in Oxyz are:

$$\varphi_1 = \frac{d^2 w_1}{d\sigma^2} = \frac{d}{d\sigma} \frac{\Omega_x}{\sqrt{1 - \frac{\Omega^2}{c^2}}} =$$

$$\frac{d}{dt}\left[\frac{\Omega_x}{\sqrt{1 - \frac{\Omega^2}{c^2}}}\right]\frac{dt}{d\sigma} = \frac{\frac{d\Omega_x}{dt}}{1 - \frac{\Omega^2}{c^2}} + \frac{\Omega_x \, \Omega \, \frac{d\Omega}{dt}}{c^2 (1 - \frac{\Omega^2}{c^2})^2},$$

$$\varphi_2 = \frac{d^2 w_2}{d\sigma^2} = \frac{d}{d\sigma} \frac{\Omega_y}{\sqrt{1 - \frac{\Omega^2}{c^2}}} =$$

$$\frac{d}{dt}\left[\frac{\Omega_y}{\sqrt{1 - \frac{\Omega^2}{c^2}}}\right]\frac{dt}{d\sigma} = \frac{\frac{d\Omega_y}{dt}}{1 - \frac{\Omega^2}{c^2}} + \frac{\Omega_y \, \Omega \, \frac{d\Omega}{dt}}{c^2 (1 - \frac{\Omega^2}{c^2})^2},$$

(A.82)

$$\varphi_3 = \frac{d^2 w_3}{d\sigma^2} = \frac{d}{d\sigma} \frac{\Omega_z}{\sqrt{1 - \frac{\Omega^2}{c^2}}} =$$

$$\frac{d}{dt}\left[\frac{\Omega_z}{\sqrt{1 - \frac{\Omega^2}{c^2}}}\right]\frac{dt}{d\sigma} = \frac{\frac{d\Omega_z}{dt}}{1 - \frac{\Omega^2}{c^2}} + \frac{\Omega_z \, \Omega \, \frac{d\Omega}{dt}}{c^2 (1 - \frac{\Omega^2}{c^2})^2},$$

$$\varphi_4 = \frac{d^2 w_4}{d\sigma^2} = \frac{d}{d\sigma} \frac{ic}{\sqrt{1 - \frac{\Omega^2}{c^2}}} =$$

$$\frac{d}{dt} \left[\frac{ic}{\sqrt{1 - \frac{\Omega^2}{c^2}}} \right] \frac{dt}{d\sigma} = \frac{i\,\Omega\,\frac{d\Omega}{dt}}{c\left(1 - \frac{\Omega^2}{c^2}\right)^2}$$

and those of $\bar{\phi}$ in $O_1x_1y_1z_1$ are:

$$\phi_1 = \frac{d^2 u_1}{d\sigma^2} = \frac{d}{d\sigma} \frac{\Omega_{x_1}}{\sqrt{1 - \frac{\Omega_1^2}{c^2}}} =$$

$$\frac{\frac{d\Omega_{x_1}}{dt_1}}{1 - \frac{\Omega_1^2}{c^2}} + \frac{\Omega_{x_1}\,\Omega_1\,\frac{d\Omega_1}{dt_1}}{c^2\left(1 - \frac{\Omega_1^2}{c^2}\right)^2},$$

$$\phi_2 = \frac{d^2 u_2}{d\sigma^2} = \frac{d}{d\sigma} \frac{\Omega_{y_1}}{\sqrt{1 - \frac{\Omega_1^2}{c^2}}} =$$

$$\frac{\frac{d\Omega_{y_1}}{dt_1}}{1 - \frac{\Omega_1^2}{c^2}} + \frac{\Omega_{y_1}\,\Omega_1\,\frac{d\Omega_1}{dt_1}}{c^2\left(1 - \frac{\Omega_1^2}{c^2}\right)^2}, \qquad \text{(A.83)}$$

$$\phi_3 = \frac{d^2u_3}{d\sigma^2} = \frac{d}{d\sigma}\frac{\Omega_{z_1}}{\sqrt{1-\frac{\Omega_1^2}{c^2}}} =$$

$$\frac{\frac{d\Omega_{z_1}}{dt_1}}{1-\frac{\Omega_1^2}{c^2}} + \frac{\Omega_{z_1}\ \Omega_1\ \frac{d\Omega_1}{dt_1}}{c^2(1-\frac{\Omega_1^2}{c^2})^2},$$

$$\phi_4 = \frac{d^2u_4}{d\sigma^2} = \frac{d}{d\sigma}\frac{ic}{1-\frac{\Omega_1^2}{c^2}} = \frac{i\ \Omega_1\ \frac{d\Omega_1}{dt_1}}{c(1-\frac{\Omega_1^2}{c^2})^2}.$$

It is interesting to note that from eqns. (A.75) and (A.76) it follows that:

$$\frac{d}{d\sigma}(q^2) = -\frac{d}{d\sigma}(c^2) = 0,$$
$$\frac{d}{d\sigma}(Q^2) = -\frac{d}{d\sigma}(c^2) = 0, \qquad (A.84)$$

which amounts to

$$\bar{q}\cdot\frac{d\bar{q}}{d\sigma} = \sum_{s=1}^{4}\frac{dw_s}{d\sigma}\frac{d^2w_s}{d\sigma^2} = 0 \qquad (A.85)$$

and

$$\bar{Q}\cdot\frac{d\bar{Q}}{d\sigma} = \sum_{s=1}^{4}\frac{du_s}{d\sigma}\frac{d^2u_s}{d\sigma^2} = 0. \qquad (A.86)$$

If one extends the notion of orthogonality, the velocity and acceleration 4-vectors are orthogonal in Minkowski space.

A.II.1.8. The momentum 4-vector. Consider again the same moving body P, treated as a material point, and let m_o be its mass measured by an observer attached to P, i.e., to the frame O'x'y'z' which is P's proper frame. The quantity m_o is referred to as the proper mass of the material point P.

The momentum 4-vector or the 4-momentum of body P with respect to frame Oxyz is defined as follows:

$$\bar{h} = m_o\bar{q}, \tag{A.87}$$

and the 4-momentum of the same body relative to the frame $O_1x_1y_1z_1$ is the 4-vector:

$$\bar{H} = m_o\ \bar{Q}. \tag{A.88}$$

Eqns. (A.68) give the components of the 4-vector $\bar{h}$ in Oxyz:

$$\begin{aligned}
h_1 &= m_oq_1 = m_o\frac{dw_1}{d\sigma} = m_o\frac{dx}{d\sigma},\\
h_2 &= m_oq_2 = m_o\frac{dw_2}{d\sigma} = m_o\frac{dy}{d\sigma},\\
h_3 &= m_oq_3 = m_o\frac{dw_3}{d\sigma} = m_o\frac{dz}{d\sigma},\\
h_4 &= m_oq_4 = m_o\frac{dw_4}{d\sigma} = m_o\frac{dt}{d\sigma}\,ic\,;
\end{aligned} \tag{A.89}$$

while eqns. (A.70) give those of the 4-vector $\bar{H}$ in $O_1x_1y_1z_1$:

$$H_1 = m_o Q_1 = m_o \frac{du_1}{d\sigma} = m_o \frac{dx_1}{d\sigma},$$

$$H_2 = m_o Q_2 = m_o \frac{du_2}{d\sigma} = m_o \frac{dy_1}{d\sigma},$$

$$H_3 = m_o Q_3 = m_o \frac{du_3}{d\sigma} = m_o \frac{dz_1}{d\sigma}, \qquad (A.90)$$

$$H_4 = m_o Q_4 = m_o \frac{du_4}{d\sigma} = m_o \frac{dt_1}{d\sigma}\, ic.$$

Substitution of eqns. (A.71) and (A.87) and (A.88) leads to the conclusion that:

$$h_k = \sum_{s=1}^{4} \alpha_{sk} H_s, \qquad H_k = \sum_{s=1}^{4} \nu_{sk} h_s; \qquad k = 1,2,3,4, \qquad (A.91)$$

which prove the 4-vector character of $\bar{h}$ and $\bar{H}$, because α_{sk} and ν_{sk} are the coefficients given by the matrices (A.24) and (A.25).

It is also readily seen, as a direct consequence of eqns. (A.75) and (A.76), that:

$$h^2 = \sum_{s=1}^{4} h_s^2 = - m_o^2 c^2, \qquad (A.92)$$

$$H^2 = \sum_{s=1}^{4} H_s^2 = - m_o^2 c^2 .$$

Therefore the square of 4-momentum is an invariant under change of reference frame.

On the other hand, eqn. (A.61) together with eqns. (A.89) yields:

$$\begin{aligned}
h_1 &= m_o \frac{dx}{dt} \cdot \frac{dt}{d\sigma} = \frac{m_o}{\sqrt{1 - \frac{\Omega^2}{c^2}}} \frac{dx}{dt} , \\
h_2 &= m_o \frac{dy}{dt} \cdot \frac{dt}{d\sigma} = \frac{m_o}{\sqrt{1 - \frac{\Omega^2}{c^2}}} \frac{dy}{dt} , \\
h_3 &= m_o \frac{dz}{dt} \cdot \frac{dt}{d\sigma} = \frac{m_o}{\sqrt{1 - \frac{\Omega^2}{c^2}}} \frac{dz}{dt} , \\
h_4 &= m_o \, ic \frac{dt}{d\sigma} = \frac{m_o}{\sqrt{1 - \frac{\Omega^2}{c^2}}} ic .
\end{aligned} \tag{A.93}$$

Likewise, eqns. (A.73) together with (A.90) yields:

$$\begin{aligned}
H_1 &= m_o \frac{dx_1}{dt_1} \cdot \frac{dt_1}{d\sigma} = \frac{m_o}{\sqrt{1 - \frac{\Omega_1^2}{c^2}}} \frac{dx_1}{dt_1} , \\
H_2 &= m_o \frac{dy_1}{dt_1} \cdot \frac{dt_1}{d\sigma} = \frac{m_o}{\sqrt{1 - \frac{\Omega_1^2}{c^2}}} \frac{dy_1}{dt_1} ,
\end{aligned} \tag{A.94}$$

$$H_3 = m_o \frac{dz_1}{dt_1} \cdot \frac{dt_1}{d\sigma} = \frac{m_o}{\sqrt{1 - \frac{\Omega_1^2}{c^2}}} \frac{dz_1}{dt_1}$$

$$H_4 = m_o \; ic \frac{dt_1}{d\sigma} = \frac{m_o}{\sqrt{1 - \frac{\Omega_1^2}{c^2}}} \; ic.$$

Consider now the quantity:

$$m = \frac{m_o}{\sqrt{1 - \frac{\Omega^2}{c^2}}}, \tag{A.95}$$

as the mass of body P in the frame Oxyz and the quantity:

$$m_1 = \frac{m_o}{\sqrt{1 - \frac{\Omega_1^2}{c^2}}}, \tag{A.96}$$

the mass of the same body P in the frame $O_1x_1y_1z_1$. Then h_1, h_2 and h_3 are the components of the momentum $\bar{p}$ of the moving body P in Oxyz and therefore:

$$\bar{p} = m \, \bar{\Omega}, \tag{A.97}$$

while H_1, H_2, H_3 are the components of momentum $\bar{p}_1$ of the same body P in $O_1x_1y_1z_1$, so that

$$\bar{p}_1 = m_1 \, \bar{\Omega}_1. \tag{A.98}$$

The dependence of mass on the frame in which it is considered has to be interpreted as follows: the inertia that a mass expresses appears as dependent on the reference frame employed, i.e., dependent on the relation between the point studied (whose mass is considered) and the observer that finds out about the mass in question.

Moreover, one can notice that if a body P moves at a speed that, relative to a frame Oxyz tends to the speed of light ($\Omega \to c$), then the inertia of P in that frame tends to infinity. This finding imposes the conclusion that the speed of light cannot be exceeded if the model implied by the theory of relativity is accepted.

A.II.1.9. The 4-force. The 4-force applied to a material point P is defined as the rate of variation of the 4-momentum of point P with respect to the proper time of P. Thus in frame Oxyz the 4-force applied to the material point P is:

$$\bar{l} = \frac{d\bar{h}}{d\sigma}, \qquad \text{(A.99)}$$

and in frame $O_1x_1y_1z_1$, the 4-force on the same point P is:

$$\bar{L} = \frac{d\bar{H}}{d\sigma}. \qquad \text{(A.100)}$$

Eqns. (A.89), (A.93) and (A.95) yield immediately the components of the 4-force:

$$l_1 = \frac{dh_1}{d\sigma} = m_o \frac{d^2 w_1}{d\sigma^2} = \frac{dh_1}{dt} \cdot \frac{dt}{d\sigma} =$$

$$\frac{d}{dt} \frac{m_o \Omega_x}{\sqrt{1 - \frac{\Omega^2}{c^2}}} \frac{dt}{d\sigma} = m \left[\frac{\Omega_x}{\sqrt{1 - \frac{\Omega^2}{c^2}}} + \frac{\Omega_x \Omega \frac{d\Omega}{dt}}{c^2 \sqrt{(1 - \frac{\Omega^2}{c^2})^3}} \right],$$

$$l_2 = \frac{dh_2}{d\sigma} = m_o \frac{d^2 w_2}{d\sigma^2} = \frac{dh_2}{dt} \cdot \frac{dt}{d\sigma} =$$

$$\frac{d}{dt} \left[\frac{m_o \Omega_y}{\sqrt{1 - \frac{\Omega^2}{c^2}}} \right] \frac{dt}{d\sigma} = m \left[\frac{\Omega_y}{\sqrt{1 - \frac{\Omega^2}{c^2}}} + \frac{\Omega_y \Omega \frac{d\Omega}{dt}}{c^2 \sqrt{(1 - \frac{\Omega^2}{c^2})^3}} \right],$$

$$l_3 = \frac{dh_3}{d\sigma} = m_o \frac{d^2 w_3}{d\sigma^2} = \frac{dh_3}{dt} \cdot \frac{dt}{d\sigma} = \qquad (A.101)$$

$$\frac{d}{dt} \frac{m_o \Omega_z}{\sqrt{1 - \frac{\Omega^2}{c^2}}} \frac{dt}{d\sigma} = m \left[\frac{\Omega_z}{\sqrt{1 - \frac{\Omega^2}{c^2}}} + \frac{\Omega_z \Omega \frac{d\Omega}{dt}}{c^2 \sqrt{(1 - \frac{\Omega^2}{c^2})^3}} \right],$$

$$l_4 = \frac{dh_4}{d\sigma} = m_o \frac{dw_4}{d\sigma^2} = \frac{dh_4}{dt} \cdot \frac{dt}{d\sigma} =$$

$$\frac{d}{dt} \left[\frac{m_o i c}{\sqrt{1 - \frac{\Omega^2}{c^2}}} \right] \frac{dt}{d\sigma} = m \frac{i \Omega \frac{d\Omega}{dt}}{c \sqrt{(1 - \frac{\Omega^2}{c^2})^3}}.$$

And similarly eqns. (A.90), (A.94) and (A.96) give the components of the 4-force $\bar{L}$:

$$L_1 = \frac{dH_1}{d\sigma} = m_o \frac{d^2u_1}{d\sigma^2} = \frac{dH_1}{dt_1} \cdot \frac{dt_1}{d\sigma} =$$

$$\frac{d}{dt_1}\left[\frac{m_o\ \Omega_{x_1}}{\sqrt{1-\frac{\Omega_1^2}{c^2}}}\right]\frac{dt_1}{d\sigma} = m_1\left[\frac{\Omega_{x_1}}{\sqrt{1-\frac{\Omega^2}{c^2}}} + \frac{\Omega_{x_1}\ \Omega_1\ \frac{d\Omega_1}{dt_1}}{c^2\sqrt{(1-\frac{\Omega_1^2}{c^2})^3}}\right],$$

$$L_2 = \frac{dH_2}{d\sigma} = m_o \frac{d^2u_2}{d\sigma^2} = \frac{dH_2}{dt_1} \cdot \frac{dt_1}{d\sigma} =$$

$$\frac{d}{dt_1}\left[\frac{m_o\ \Omega_{y_1}}{\sqrt{1-\frac{\Omega_1^2}{c^2}}}\right]\frac{dt_1}{d\sigma} = m_1\left[\frac{\Omega_{y_1}}{\sqrt{1-\frac{\Omega_1^2}{c^2}}} + \frac{\Omega_{y_1}\ \Omega_1\ \frac{d\Omega_1}{dt_1}}{c^2\sqrt{(1-\frac{\Omega_1^2}{c^2})^3}}\right],$$

$$L_3 = \frac{dH_3}{d\sigma} = m_o \frac{d^2u_3}{d\sigma^2} = \frac{dH_3}{dt_1} \cdot \frac{dt_1}{d\sigma} = \qquad (A.102)$$

$$\frac{d}{dt_1}\left[\frac{m_o\ \Omega_{z_1}}{\sqrt{1-\frac{\Omega_1^2}{c^2}}}\right]\frac{dt_1}{d\sigma} = m_1\left[\frac{\Omega_{z_1}}{\sqrt{1-\frac{\Omega_1^2}{c^2}}} + \frac{\Omega_{z_1}\ \Omega_1\ \frac{d\Omega_1}{dt_1}}{c^2\sqrt{(1-\frac{\Omega_1^2}{c^2})^3}}\right],$$

$$L_4 = \frac{dH_4}{d\sigma} = m_o \frac{d^2u_4}{d\sigma^2} = \frac{dH_4}{dt_1} \cdot \frac{dt_1}{d\sigma} =$$

$$\frac{d}{dt_1}\left[\frac{m_o\ i\ c}{\sqrt{1-\frac{\Omega_1^2}{c^2}}}\right]\frac{dt_1}{d\sigma} = m_1 \frac{i\ \Omega_1\ \frac{d\Omega_1}{dt_1}}{c\sqrt{(1-\frac{\Omega^2}{c^2})^3}}.$$

Eqns. (A.91) allow one to infer that:

$$l_k = \sum_{s=1}^{4} \alpha_{sk} L_s; \; L_k = \sum_{s=1}^{4} \nu_{sk} l_s ; \qquad (A.103)$$

$$k = 1,2,3,4;$$

which shows the 4-vector character of the 4-force.

A.II.1.10. <u>The fundamental law of dynamics and the energy of a particle in relativistic dynamics</u>. The force that acts upon a material point P, i.e., the rate of variation of momentum, is, in Oxyz:

$$\bar{F} = \frac{d\bar{p}}{dt} = \frac{d}{dt} (m \bar{\Omega}) \qquad (A.104)$$

and in $O_1x_1y_1z_1$:

$$\bar{F}_1 = \frac{d\bar{p}_1}{dt_1} = \frac{d}{dt_1} (m_1 \bar{\Omega}_1). \qquad (A.105)$$

Eqn. (A.104) represents the fundamental law of dynamics in the frame Oxyz, while (A.105) represents the same law in the frame $O_1x_1y_1z_1$.

Taking into account eqns. (A.85)-(A.88), (A.99) and (A.100), one gets:

$$\bar{h} \cdot \bar{l} = \sum_{s=1}^{4} h_s l_s = 0;$$

$$\bar{H} \cdot \bar{L} = \sum_{s=1}^{4} H_s L_s = 0. \qquad (A.106)$$

On the other hand, eqns. (A.93) and (A.101) on the one hand and (A.94) and (A.102) on the other yield, after simple calculations:

$$\sum_{s=1}^{4} h_s l_s = 0; \quad \sum_{s=1}^{4} H_s L_s = 0; \qquad \text{(A.106')}$$

which are evidently perfectly equivalent to:

$$\Omega_x \frac{d}{dt}(m\,\Omega_x) + \Omega_y \frac{d}{dt}(m\,\Omega_y) + \Omega_z \frac{d}{dt}(m\,\Omega_z) = \frac{d}{dt}(m\,c^2)$$

$$\Omega_{x_1} \frac{d}{dt_1}(m_1\,\Omega_{x_1}) + \Omega_{y_1} \frac{d}{dt_1}(m_1\,\Omega_{y_1}) + \Omega_{z_1} \frac{d}{dt_1}(m_1\,\Omega_{z_1}) = \frac{d}{dt_1}(m_1\,c^2), \qquad \text{(A.107)}$$

meaning that:

$$\bar{\Omega}\cdot\frac{d}{dt}(m\,\bar{\Omega}) = \frac{d}{dt}(m\,c^2) \qquad \text{(A.108)}$$

and

$$\bar{\Omega}_1\cdot\frac{d}{dt_1}(m_1\,\bar{\Omega}_1) = \frac{d}{dt_1}(m_1\,c^2). \qquad \text{(A.109)}$$

Now, according to (A.104) and (A.105) one gets:

$$\bar{F}\cdot\bar{\Omega} = \frac{d}{dt}(m\,c^2); \quad \bar{F}_1\,\bar{\Omega}_1 = \frac{d}{dt_1}(m_1\,c^2). \qquad \text{(A.110)}$$

However, the scalar product of force and velocity is the rate of variation (the derivative with respect to time) of energy. Thus:

$$E = m\,c^2 = \frac{m_o\,c^2}{\sqrt{1 - \frac{\Omega^2}{c^2}}} \tag{A.111}$$

is the energy of the particle for an observer attached to the frame Oxyz and:

$$E_1 = m_1\,c^2 = \frac{m_o\,c^2}{\sqrt{1 - \frac{\Omega_1^2}{c^2}}} \tag{A.112}$$

is the energy for an observer attached to frame $O_1x_1y_1z_1$.

As to the components F_x, F_y, F_z of force $\bar{F}$ and F_{x_1}, F_{y_1}, F_{z_1} of $\bar{F}_1$ and the energies E and E_1, they transform on transfer from frame Oxyz to frame $O_1x_1y_1z_1$ (and conversely) as follows:

$$F_x = \frac{F_{x_1} - \frac{v}{c^2}\,\bar{F}_1 \cdot \bar{\Omega}_1}{1 - \frac{v}{c^2}\,\Omega_{x_1}}; \quad F_y = \frac{F_{y_1}\sqrt{1 - \frac{v^2}{c^2}}}{1 - \frac{v}{c^2}\,\Omega_{x_1}};$$

$$F_z = \frac{F_{z_1}\sqrt{1 - \frac{v^2}{c^2}}}{1 - \frac{v}{c^2}\,\Omega_{x_1}}; \quad E = E_1\,\frac{1 - \frac{v}{c^2}\,\Omega_{x_1}}{\sqrt{1 - \frac{v^2}{c^2}}}; \tag{A.113}$$

and

$$F_{x_1} = \frac{F_x + \frac{v}{c^2}\,\bar{F} \cdot \bar{\Omega}}{1 + \frac{v}{c^2}\,\Omega_x}; \quad F_{y_1} = \frac{F_y\sqrt{1 - \frac{v^2}{c^2}}}{1 + \frac{v}{c^2}\,\Omega_x}; \tag{A.114}$$

$$F_{z_1} = \frac{F_z \sqrt{1 - \frac{v^2}{c^2}}}{1 + \frac{v}{c^2}\Omega_x};\quad E_1 = E\,\frac{1 + \frac{v}{c^2}\Omega_x}{\sqrt{1 - \frac{v^2}{c^2}}};$$

which are direct consequences of the Lorentz-Einstein transformations and of the expressions for force and energy.

A.II.1.11. <u>The momentum-energy vector and the properties of its components</u>. The momentum 4-vectors may be represented as follows:

$$\bar{h} = \bar{h}\,(m\,\bar{\Omega},\ \frac{i}{c}\,E);\quad \bar{H} = \bar{H}\,(m_1\,\bar{\Omega}_1,\ \frac{i}{c}\,E_1); \qquad (A.115)$$

due to eqns. (A.93), (A.94), (A.97), (A.98), (A.111) and (A.112) which also yield:

$$h^2 = p^2 - \frac{E^2}{c^2};\quad H^2 = p_1^2 - \frac{E_1^2}{c^2}. \qquad (A.116)$$

Now, if eqn. (A.92) is considered, it means that:

$$E = c\sqrt{p^2 + m_o^2\,c^2};\quad E_1 = c\sqrt{p_1^2 + m_o^2\,c^2}. \qquad (A.117)$$

The $E_o = m_o\,c^2$ is referred to as the <u>rest energy</u> because it is obtained from the expression for E, by setting $\Omega = 0$, or from the expression of E_1 when $\Omega_1 = 0$.

Therefore E_o is the energy of the material point P in its proper frame.

Consequently one can still write:

$$E = c\sqrt{p^2 + m_o\,E_o};\quad E_1 = c\sqrt{p_1^2 + m_o\,E_o}. \qquad (A.118)$$

To establish an important feature of the relativistic energy one expands the functions $E = E\ (\Omega)$ and $E_1 = E_1\ (\Omega_1)$ in Taylor series:

$$E = m\ c^2 = \frac{m_o\ c^2}{\sqrt{1 - \frac{\Omega^2}{c^2}}} = m_o\ c^2 +$$

$$\frac{1}{2}\ m_o\ \Omega^2 + \frac{3}{8}\ m_o\ \frac{\Omega^4}{c^2} + \ldots,$$

(A.119)

$$E_1 = m_1\ c^2 = \frac{m_o\ c^2}{\sqrt{1 - \frac{\Omega_1^2}{c^2}}} = m_o\ c^2 +$$

$$\frac{1}{2}\ m_o\ \Omega_1^2 + \frac{3}{8}\ m_o\ \frac{\Omega_1^4}{c^2} + \ldots,$$

The terms involving the inverse of the speed of light with powers higher than unity can be discarded and so one gets, to a good approximation:

$$E \cong m_o\ c^2 + \frac{1}{2}\ m_o\ \Omega^2;\ E_1 \cong m_o\ c^2 + \frac{1}{2}\ m_o\ \Omega_1^2. \qquad \text{(A.120)}$$

The quantity

$$T = \frac{1}{2}\ m_o\ \Omega^2 \qquad \text{(A.121)}$$

is obviously the kinetic energy of the point P in the frame Oxyz and

$$T_1 = \frac{1}{2} m_o \Omega_1^2 \qquad (A.122)$$

is the kinetic energy of the point P in the frame $O_1x_1y_1z_1$.

Eqns. (A.120) mean in fact that $E = E_o + T$ and $E_1 = E_o + T_1$.

E_o is an invariant, that is an intrinsic characteristic of the material point, while the kinetic energy depends on the velocity of the material point relative to the frame. E_o is the internal energy, also referred to as the rest energy.

A.II.1.12. The conservative case. Potential energy. The force $\bar{F}$ introduced through eqn. (A.104) is called a conservative force provided that there exists a scalar function $N = N(x,y,z)$, which depends on the coordinates x, y, z of the material point P in the frame Oxyz, such that:

$$\bar{F} = \bar{\nabla} N. \qquad (A.123)$$

This comes to the same thing as:

$$F_x = \frac{\partial N}{\partial x}; \; F_y = \frac{\partial N}{\partial y}; \; F_z = \frac{\partial N}{\partial z}. \qquad (A.124)$$

Then, because $\bar{F} \cdot \bar{\Omega} = \frac{dE}{dt}$, one may write:

$$\frac{\partial N}{\partial x} \Omega_x + \frac{\partial N}{\partial y} \Omega_y + \frac{\partial N}{\partial z} \Omega_z = \frac{dE}{dt}. \qquad (A.125)$$

And, from $\Omega_x\, dt = dx$, $\Omega_y\, dt = dy$, $\Omega_z\, dt = dz$, it follows that:

$$\frac{\partial N}{\partial x}\,dx + \frac{\partial N}{\partial y}\,dy + \frac{\partial N}{\partial z}\,dz = dN = dE \qquad (A.126)$$

The potential energy is

$$W = -\,N, \qquad (A.127)$$

which means that eqn. (A.126) is identical with $d(E + W) = 0$, i.e.,

$$E + W = E_T = \text{constant}. \qquad (A.128)$$

This is the law of conservation of mechanical energy, in which E_T stands for the total energy of the material point.

According to eqn. (A.118) one can utilize the form:

$$E_T = c\sqrt{p^2 + m_o E_o} + W. \qquad (A.129)$$

If the force $\bar{F}$ is conservative, then the force $\bar{F}_1$ which, for an observer in the reference frame $O_1x_1y_1z_1$, acts upon the same point P is given by

$$\bar{F}_1 = \frac{\partial N_1}{\partial x_1}\,\bar{i}_1 + \frac{\partial N_1}{\partial y_1}\,\bar{j}_1 + \frac{\partial N_1}{\partial z_1}\,\bar{k}_1, \qquad (A.130)$$

where x_1, y_1, z_1 are the coordinates of P; $\bar{i}_1$, $\bar{j}_1$, $\bar{k}_1$ are the unit vectors along the axes O_1x_1, O_1y_1, O_1z_1, respectively; and $N_1 = N_1\,(x_1, y_1, z_1, t_1)$ is a function expressing a scalar field which depends on time.

One can see that the partial derivatives of this function must satisfy the conditions:

$$\frac{\dfrac{\partial N_1}{\partial x_1}}{\sqrt{1-\dfrac{\Omega_1^2}{c^2}}} = \frac{\dfrac{\partial N}{\partial x} + \dfrac{v}{c^2}\,\dfrac{\dfrac{d}{dt}\left(\frac{1}{2}\, m_o\, \Omega^2\right)}{\left(1-\dfrac{\Omega^2}{c^2}\right)^{3/2}}}{\sqrt{1-\dfrac{v^2}{c^2}}\;\sqrt{1-\dfrac{\Omega^2}{c^2}}};$$

$$\frac{\dfrac{\partial N_1}{\partial y_1}}{\sqrt{1-\dfrac{\Omega_1^2}{c^2}}} = \frac{\dfrac{\partial N}{\partial y}}{\sqrt{1-\dfrac{\Omega^2}{c^2}}};\quad \frac{\dfrac{\partial N_1}{\partial z_1}}{\sqrt{1-\dfrac{\Omega_1^2}{c^2}}} = \frac{\dfrac{\partial N}{\partial z}}{\sqrt{1-\dfrac{\Omega^2}{c^2}}}, \tag{A.131}$$

that have been established using also eqns. (A.113), (A.114).

It is obvious that, in Newtonian mechanics, $v\, c^{-2}$ 0; $\Omega^2\, c^{-2} \cong 0$; $\Omega_1^2\, c^{-2} \cong 0$, so that

$$\frac{\partial N_1}{\partial x_1} = \frac{\partial N}{\partial x};\quad \frac{\partial N_1}{\partial y_1} = \frac{\partial N}{\partial y};\quad \frac{\partial N_1}{\partial z_1} = \frac{\partial N}{\partial z}; \tag{A.131'}$$

on the other hand, one should note that $N_1 = N_1\,(x_1, y_1, z_1, t_1)$ will continue to depend on time, which gives the field in $O_1x_1y_1z_1$ a varying character.

A.II.1.13. The mass defect. The relativistic expression for energy emphasizes an interesting phenomenon. The presentation utilizes the frame Oxyz but, the transformation relations allow transfer to $O_1x_1y_1z_1$.

Consider n material particles M_s whose masses at a certain time are m_s, where s = 1,2,...,n. It is assumed that the particles form an isolated material system, i.e., one whose total energy is conserved. Consider

W_{ij}, the potential energy of particle M_i resulting from its interaction with M_j, at the same moment t when the masses m_s are considered. At a time $t' > t$ the masses are m'_s and the potential energies of particles M_i, generated by interactions with M_j, are W'_{ij}. Obviously, $i = 1,2,\ldots,n$; $j = 1,2,\ldots,n$, and $i = j$ implies $W_{ij} = W'_{ij} = 0$.

The law of conservation of energy yields:

$$\sum_{s=1}^{n} m_s c^2 + \sum_{i=1}^{n} \sum_{j=1}^{n} W_{ij} = E_T = \sum_{s=1}^{n} m'_s c^2 + \sum_{i=1}^{n} \sum_{j=1}^{n} W'_{ij}. \tag{A.132}$$

With the notations:

$$m'_s - m_s = \Delta m_s; \quad \sum_{s=1}^{n} \Delta m_s = \Delta m;$$
$$W'_{ij} - W_{ij} = \Delta W_{ij}; \quad \sum_{i=1}^{n} \sum_{j=1}^{n} \Delta W_{ij} = \Delta W; \tag{A.133}$$

it follows that

$$c^2 \Delta m = - \Delta W. \tag{A.134}$$

Consider now a body K whose mass in its proper frame is m_o.

Then the rest energy, the internal energy of the body, is $E_o = m_o c^2$.

Suppose now that this body blows up spontaneously,

i.e. , without outside interference. It divides into n parts K_s of masses m_s and velocities Ω_s $(s = 1,2,..,n)$ relative to the proper frame of body K. The whole phenomenon is studied relative to the proper frame of K.

If the proper masses of particles are K_s and m_{os}, then:

$$m_s = \frac{m_{os}}{\sqrt{1 - \frac{\Omega_s^2}{c^2}}} \tag{A.135}$$

and the conservation of energy requires that:

$$m_o c^2 = \sum_{s=1}^{n} m_s c^2 = \sum_{s=1}^{n} \frac{m_{os} c^2}{\sqrt{1 - \frac{\Omega_s^2}{c^2}}}. \tag{A.136}$$

If particles K_s give up their entire kinetic energy to their surroundings, each retains $E_{os} = m_{os} c^2$. Thus the surroundings receive the energy:

$$\Delta E = m_o c^2 - \sum_{s=1}^{n} m_{os} c^2 = \left[\sum_{s=1}^{n} m_{os} \left(\frac{1}{\sqrt{1 - \frac{\Omega_1^2}{c^2}}} - 1 \right) \right] c^2. \tag{A.137}$$

Obviously $\Delta E > 0$ and $\Delta m_o = m_o - \sum_{s=1}^{n} m_{os} > 0$.

The eqn. (A.137) is the well known relation $\Delta E =$

$c^2 \Delta m_o$, Δm_o being the mass defect.

A body that disintegrates spontaneously is unstable. The instability of the body occurs if $m_o > \sum_{s=1}^{n} m_{os}$.

If however $m_o < \sum_{s=1}^{n} m_{os}$, the body is stable.

Consider now a stable body K_1 whose mass is m_{o_1}.

The phenomenon is studied against the proper frame of K_1, in which K_1's energy is $E_{o_1} = m_{o_1} c^2$. It is supposed that by some process the internal energy of K_1 increases without modifying its kinetic energy. Let E' be the energy gain. Then K_1's mass becomes:

$$\mu_{o_1} = \frac{E_{o_1} + E'}{c^2} = m_{o_1} + \frac{E'}{c^2}. \tag{A.138}$$

If on reaching the rest mass μ_{o_1} the body turns unstable and the disintegration studied above occurs, then $E' \geqslant (\mu_{o_1} - m_{o_1})c^2$ is the energy required to disintegrate the stable body.

A.II.1.14. The relativistic study of collisions. Consider two material particles M and N of rest masses m_o and μ_o, respectively. Let us study their collision and try to determine the variation of velocity caused by the impact, assuming that their motion is referred to a certain frame Oxyz. The same phenomenon may be studied in $O_1x_1y_1z_1$, inertial with respect to the frame Oxyz, if one utilizes the transformation discussed above.

Assume $\bar{\Omega}$ was the velocity of M and $\bar{v}$ the velocity of N just before collision.

Immediately after collision these vectors become $\bar{\Omega}'$ and $\bar{v}'$, respectively.

In effect, the interaction takes place in a

direction perpendicular to the common tangent plane of the two particles, the two particles being, after all, bodies whose substance is inside a nonzero volume, however small this volume might be (Fig. A-1).

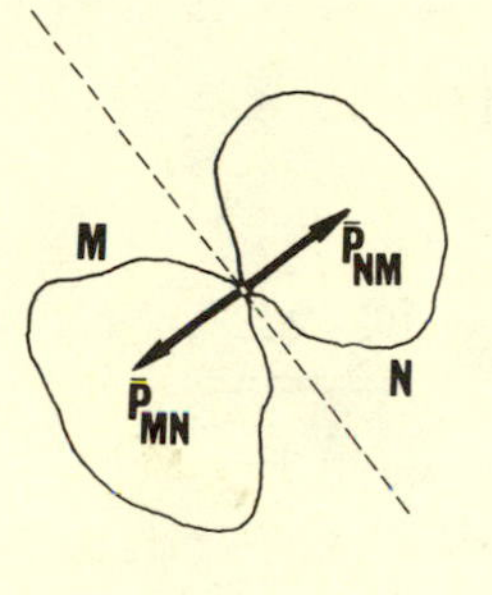

Fig. A-1.

The percussion vectors $\bar{P}_{NM}$ and $\bar{P}_{MN}$ indicate the jump in the momentum of the particles. These sudden variations occur however only along the interaction direction.

The discussion is restricted only to the case when $\bar{\Omega}$ and $\bar{v}$ are collinear and along the direction of interaction. This is a particular case, but it contains all the relevant findings required here, because if $\bar{\Omega}$ and $\bar{v}$ had some components perpendicular to the direction of interaction these would not be affected by collision.

Thus $\bar{\Omega}$, $\bar{v}$, $\bar{\Omega}'$, and $\bar{v}'$ are collinear vectors. The collision is assumed to be characterized by a coefficient of restitution k, given by:

$$k = \frac{\Omega' - v'}{v - \Omega}. \tag{A.139}$$

The conservation of the momentum 4-vector involves conservation of both the momentum and the energy.

Thus, if m_o' and μ_o' denote the rest masses of points M and N, after collision one can write:

$$\frac{m_o \Omega}{\sqrt{1 - \frac{\Omega^2}{c^2}}} + \frac{\mu_o v}{\sqrt{1 - \frac{v^2}{c^2}}} = \frac{m_o' \Omega'}{\sqrt{1 - \frac{\Omega'^2}{c^2}}} + \frac{\mu_o' v'}{\sqrt{1 - \frac{v'^2}{c^2}}} \tag{A.140}$$

for the conservation of momentum and

$$\frac{m_o c^2}{\sqrt{1 - \frac{\Omega^2}{c^2}}} + \frac{\mu_o c^2}{\sqrt{1 - \frac{v^2}{c^2}}} = \frac{m_o' c^2}{\sqrt{1 - \frac{\Omega'^2}{c^2}}} + \frac{\mu_o' c^2}{\sqrt{1 - \frac{v'^2}{c^2}}}. \tag{A.141}$$

for the conservation of energy.

Thus, there are three equations, (A.139), (A.140) and (A.141), with four unknowns: Ω', v', m', μ_o'. Unlike what happens in classical mechanics, the modification of rest masses introduces unknowns m' and μ', which give the problem an undetermined character.

This is natural because, depending on the deformability and physical properties of particles, they store more or less internal energy, which alters their rest mass. In fact, these properties, which determine the internal energy absorbed by each particle, also determine the value of the restitution coefficient, which depends on the energy absorbed.

A practical solution requires the (experimental) determination of the energy ratio of the two particles involved in collision.

An interesting particular case is that of an inelastic collision, when $k = 0$ and $\Omega' = v'$.

In this case, after collision the two masses remain united in one body. It is shown below that the problem of inelastic collision is determined because the bodies coalesce and the determinate, because the masses m_o' and μ_o' is not needed any longer; only the rest mass q_o of the final body is sought. Thus two unknowns m_o' and μ_o' are replaced by one, q_o.

Let q_o be the rest mass of particles which have coalesced following collision.

Eqns. (A.140) and (A.141) reduce to:

$$\frac{m_o \ \Omega}{\sqrt{1 - \frac{\Omega^2}{c^2}}} + \frac{\mu_o \ v}{\sqrt{1 - \frac{v^2}{c^2}}} = \frac{q_o \ \Omega'^2}{\sqrt{1 - \frac{\Omega^2}{c^2}}}, \qquad \text{(A.142)}$$

$$\frac{m_o \ c^2}{\sqrt{1 - \frac{\Omega^2}{c^2}}} + \frac{\mu_o \ c^2}{\sqrt{1 - \frac{v^2}{c^2}}} = \frac{q_o \ c^2}{\sqrt{1 - \frac{\Omega'^2}{c^2}}}. \qquad \text{(A.143)}$$

Substitution of (A.143) in (A.142) yields:

$$\Omega' = \frac{\dfrac{m_o \Omega}{\sqrt{1 - \dfrac{\Omega^2}{c^2}}} + \dfrac{\mu_o v}{\sqrt{1 - \dfrac{v^2}{c^2}}}}{\dfrac{m_o}{\sqrt{1 - \dfrac{\Omega^2}{c^2}}} + \dfrac{\mu_o}{\sqrt{1 - \dfrac{v^2}{c^2}}}}, \qquad \text{(A.144)}$$

or, alternatively:

$$\Omega' = \frac{m_o \Omega \sqrt{1 - \dfrac{v^2}{c^2}} + \mu_o v \sqrt{1 - \dfrac{\Omega^2}{c^2}}}{m_o \sqrt{1 - \dfrac{v^2}{c^2}} + \mu_o \sqrt{1 - \dfrac{\Omega^2}{c^2}}}. \qquad \text{(A.144)}$$

Once eqn. (A.144') is inserted into (A.143), one finds the rest mass q_o of particles which have coalesced in a plastic collision.

It is easy to see that, in case the velocities are low compared to the speed of light, that is, $\Omega^2/c^2 \cong 0$ and $v^2/c^2 \cong 0$, the equations above assume the form known from classical mechanics, which proves, in this circumstance, to be a good approximation of the reality.

It is noteworthy that, starting from the study of inelastic collision, one could obtain the Mescerski-Levi Civita equation for the relativistic case, that is, the equation of motion of a body which emits or captures a jet of particles. This study may be undertaken considering first only the capture, as an inelastic collision, then the emission as a negative capture.

A.II.1.15. Some physical interpretations and other remarks on relativistic mechanics. It is natural that a theory so comprehensive as relativity, with deep implications that affect the very background of our notions, must have far-reaching impacts upon various fields of knowledge.

Given its complexity, the physical and philosophical interpretation of this theory has not been and could not possibly be free of controversy.

For a correct understanding of nature, one should not loose sight of one fact: that any scientific theory, in which a concept is related to an ensemble of phenomena, begins by constructing a model in which Nature (or its part that has been the subject of research) is represented by a simplified scheme which approximates the reality.

On the one hand, the closer it comes to the reality, the more advanced the model is; on the other hand, any model, any general image of the environment and of the phenomena observed represents a synthesis, an abstraction of experimental findings and of measurements carried out on parameters whose variation characterizes the ensemble of phenomena. Such syntheses, abstractions and generalizations, which are always needed in science, have led to the basic notions of space and time.

Thus, one can do modelling only on what one can actually observe and measure.

However, all processes of observation and measurement require the existence of an agent that carries information from the phenomenon in question to the site where it is studied (some measuring equipment, observer etc.). This information carrier makes possible the study of the states of the universe, their succession (by one of the syntheses mentioned), i.e., the notion of time. The information carrying agent is, conspicuously, light (electromagnetic waves), since when the generalization and abstraction of data noted in the process of modelling (syntheses) is thus expressed in a quantitative form, one prefers the <u>most rapid</u> information carrier.

The Lorentz relations, which are the generating element of the theory of relativity, do nothing but take account of the information carrier which is ignored in classical pre-relativistic science; here "ignored" means that the information is given no role in the process of synthesising the data measured, i.e., in the structuring of a general model. Thus, in classical physics, terms such as "time", "simultaneity", "succession" are used without a profound analysis of how and who establishes the simultaneity and succession of some phenomena, in fact, of the passing of what we call time.

In this way, classical physics achieves models which are, to a large extent, abstract. An absolute time and absolute space are accepted, but even these are poorly defined.

The more factors of reality that are disregarded in making a model, the simpler, more idealized and more approximate (far from reality) is the model. Classical physics offers an approximate image of the universe, because it ignores the fact that information, which accounts for the raw material for generating general images of nature, travels by means of a carrier agent, that this transport of information is an essential phenomenon and that any form of interaction obeys these general laws of information transport (whenever it involves material systems that are far from each other). This approximate character was evidenced when the accuracy of measurements had reached an adequate level (obtained through the development of technical methods of measurement) to sense the difference between nature itself and the (classical) model.

The role of the information carrier (light) is outstandingly important when the observer and the event observed are in relative motion.

The results of the measurement are influenced by this motion (the relative motion of the phenomenon studied and the observer), because it determines how by means of light (or, more generally, of electromagnetic waves or any other phenomenon of propagation of equal speed) information regarding the quantities measured is carried.

The Lorentz transformations and, together with them, the whole theory of relativity attempt to take account of this phenomenon of information transport, of the influence it exerts, due to the transport with high but finite velocity (velocity of light), and of the relative motion of the phenomenon and observer (which implies a difference between the findings of two observers in relative motion, even when their findings regard the same events).

If things are regarded this way, the profound phenomenologic core of the theory seems to consist in emphasizing a relativity of the measurement or, more precisely, in the influence that the measurements necessarily derive from the phenomenon of information transport.

Now, because transport of information is, in its turn, directly determined by the relative state of motion, it follows that the findings are altered by the state of motion. That the information transport is determined by the relative states of motion is obvious, because one has to deal, in fact, with a composition of two motions: that of the information carrier and the relative one between the event (from which information is carried) and the observer (to whom it is carried) Or, between two different observers (because they move differently with respect to the event studied). Or, in the case where two motions are composed, both components are meaningful. It is true that the weight of one of the components (in the case under consideration, the relative motion of the event and the observer or the relative motion of the frames the observers are attached to) may eventually be disregarded, but <u>only</u> as a first approximation.

Thus, when interpreted as an introduction to the calculations and reasoning on the transport (at the speed of light, which is high but finite) of information, interaction energy etc. - a transport that influences the quantitative findings (measurements) the theory of relativity should look less embarrassing to common sense.

The contraction of space is a contraction of measured length when the segment measured and the observer who carries out the measurement participate in a relative motion. The proper length (in the frame attached to the segment) does not change.

The same holds true for the expansion of time intervals.

Now the question arises as to whether these statements reduce, in fact, the meaning of the theory of relativity only to some corrections, indeed important but not strictly related to the essence of the phenomena, corrections that may be regarded as a theory of quantitative determinations (measurements) when the results of these are transported by an information carrier.

In such a case, the theory of special relativity would do no more than draw attention to the fact that data do not come directly from the source but are carried by some agent, as well as introducing into reasoning and calculations this phenomenon of transport of information.

If so, what hides behind the way information propagates, behind the way various quantities are measured or influences are transported? In other words, what is the reality, "stripped" of all intermediate influence, of the transport factor (light or electromagnetic wave)?

It is hard to say whether or not asking such questions make sense. What is of real essence is that one can perceive, measure and consequently interpret and model Nature only on the basis of what is found by experiment, on the basis of what the information carrier brings with it from the environment. The reality is the one which is noted, the one that is received.

This is not to plead for empiricism. It is just to express the opinion that the phenomena in the universe exert a reciprocal influence which implies the existence of another phenomenon, the propagation of influence; and as a result any finding, interpretation and modelling bears the stamp of the phenomenon of transport of influence (information). This transport occurs in different ways, depending on the relative motion of information source and receiver. This means quantitatively different information for different receivers (from the viewpoint of their motion).

This is "relativity", and seen thus it should not cause anxiety, neither should it offend common sense; it is only in conflict with a certain approximate model of reality, the classical model, which accounts for only a part of reality (ignoring the transport and propagation of influence).

This classical model used to be and still is eventually identified as the Nature itself. This is certainly a misconception.

However, it seems a fact that the axiom of the constancy of the speed of light, fundamental as it is for the theory of relativity, is not entirely satisfactory for the human spirit, although it is a fact verified by experiment. The dissatisfaction, or rather the psychological hangover associated with this postulate, stems apparently from its surprising character. And this feature is generated, as we believe, by an insufficient analysis of the notion of time.

This work aims to undertake precisely such an analysis.

A.II.2. Some aspects of the theory of general relativity

Although special relativity is of major importance in the study of nature and in technical applications of high-speed motion, its approach to phenomena is peculiar: the reference frames considered are inertial. In fact the notion of inertial frame may be the object of a profound analysis. These frames were defined as those in which a point not subjected to any interaction moves rectilinearly and uniformly; it was, then, only natural to extend the knowledge of the relativistic aspect of motion to frames against which a particle not acted upon moves with a certain acceleration.

The presence of acceleration independent of the mass of the particle appears as a property of the space of such a frame, a property that may be generated either by the non-inertial motion of the reference frame or by the presence of masses which modify the character of space. Such spaces, in which every material point has an acceleration, are referred to as acceleration fields.

In the case when the acceleration field is determined only by non-inertial motion of the frame, one should observe that a suitable choice of another frame reduces to zero the acceleration field; thus, one can always find an inertial frame against which the accelerations of points not acted upon are zero. This is impossible when the acceleration field is a consequence of the presence of masses.

This is an essential difference, but not the only one, between the fields caused exclusively by the non-inertiality of the frame and those originating in the presence of masses. It should be added here that, while the acceleration fields generated by a mass tend to zero for points at infinity, the accelerations with respect to frames in non-inertial motion increases or remain unchanged (anyway they do not disappear) for points that are far from the origin of the frame.

To achieve a unified treatment of the theory of acceleration fields, A. Einstein has considered these differences and the common feature of all acceleration fields: the acceleration of one particle in the field does <u>not</u> depend on its mass.

The relativistic theory of gravitational fields,

of these acceleration fields generated either by non-inertial motion of the frame or by presence of masses, is the theory of general relativity.

To develop it, Einstein has generalized the motion of reference frame, resorting to Gaussian coordinates, has imagined the whole space filled up with some continuous "medium", the time varying independently at each point. The synchronization of clocks is no longer possible (because the relative motion of frames is not inertial), but between infinitely close points the differences in the variation of time have to be indefinitely small.

This continuous "medium", this space-frame, is not necessarily rigid but may be deformable. It is a kind of reference "mollusc" as Einstein called it. This way it was possible to state the principle of equivalence of all frames with respect to which the laws of nature have an unified form.

If u_1, u_2, u_3, u_4 are the space-time coordinates (three of space and one of time) in such a frame, then the square of an interval is expressed by:

$$ds^2 = - \sum_{i=1}^{4} \sum_{k=1}^{4} g_{ik}\, du_i\, du_k \qquad \text{(A.145)}$$

where g_{ik} depends on space and time.

This is a Riemannian metric. It is said that in the presence of gravitational fields there is a curvature of space-time.

Minkowski space is a particular case obtaining when: $g_{ik} = 1$ for $i = k$, $g_{ik} = 0$ for $i \neq k$ ($u_1 = x$, $u_2 = y$, $u_3 = z$, $u_4 = ict$).

The space-time paths in such gravitational fields

are no longer straight lines in the case of particles not acted upon by forces; they are geodetic curves specific to the space-time structure; they are characterized by the Riemannian metric (A.145).

The general study of such a space-time, with local properties variable with the energy states, is, in fact, the theory of general relativity, understood correctly by many specialists as a theory not only very complex but actually excellent with respect to efficiency and generality.

The laws of nature have the same expression in any frame; the phenomena of inertia and of gravitational attraction are integrated in the same more general frame (the equivalence principle).

However this remarkable theory is not presented here, because, in the following, we shall specialize in some aspects of mechanical motion which sometimes agree with general relativity, sometimes differ from it in approach, though without ever conflicting with general relativity; to understand what is presented further on, one needs, besides knowledge of classical mechanics, only the matter presented above.

It will be easy to see when the procedures, premises or conclusions are similar with or different from those of the theory of general relativity.

A.III. THE MAIN AIMS OF THIS WORK

It will be evident in the sequel that this work aims at going deeper into some notions and basic aspects related to several fields of physics, in an attempt to give them additional consistency, clarity and relevance.

Taking a step forward towards more adequate concepts in modelling Nature was essayed in those areas where progress was considered both needed and possible.

The concept of time was analyzed and an attempt was made to give it a rigorous foundation. Thus, the conclusion was reached that expressing the speed of light by the same constant in any inertial reference frame reflects a truth that may be reached in a different manner from the axiomatic statement. Turning the axiom that underlies the theory of special relativity into a theorem that appears as both natural and necessary brings about major implications in the understanding and interpretation of Einstein's relativity and in any other study that employs time.

An attempt at an analysis, an interpretation and a new definition of inertial and non-inertial reference frames, as well as of inertia itself, was also undertaken. In turn, this has required an investigation of the properties of space, that is of space-time aspects - since space and time exist inseparably. In modern physics, space may not be conceived apart from time, neither can time be conceived without space. Thus a definition - temporal in nature - of some properties of space and of geometric varieties has been introduced.

Besides these attempts with a more basic character, this work contains a generalization of the Levi-Civita equations describing the dynamics of a particle with variable mass due to capture or emission of substance.

Finally, at the end of the book a proposal is made towards a modification of the axiomatic basis of Newtonian mechanics, in a way that does not contradict classical axiomatics.

B. TIME

B.I. THE CONCEPT OF TIME

The understanding of the concepts of space and time has experienced extensive changes.

The evolution of attempts to define time and space, to introduce these quantities, whose importance for knowledge in general and physics in particular is obvious, in the reasoning and calculations of various sciences mirrors the evolution of scientific and philosophic thinking.

B.I.1. The evolution of the concept of time

Although in the following the existing definitions and the most important interpretations regarding time and space are assumed known, some aspects related to the understanding and utilization of the concept of time, which is the focus of our attention in this chapter, are briefly reviewed and analyzed as an introduction to this chapter.

The two quantities, space and time, which are present in most scientific reasoning, in most acts of knowing and in current observations and experiments, and which are encountered in everyday life, have long been considered and utilized as concepts whose deep meaning has not been (and could not have been) rigorously defined. Therefore, whenever quantitative expressions were needed, one usually resorted to measurements whose accuracy depended obviously on the available technology at that particular time. However, this fact of tremendous practical importance does not affect the background aspects. Measurements have been (and are) processes of comparison of measured quantities, with some others taken as standard. In other words, the practical requirements were complied with, while avoiding deeper generalizations required by theoretical concepts.

There was no model to attempt a simplified, idealized yet coherent representation of Nature.

Such a model, a consistent image underlied by rigorously defined notions was supplied for the first time by theoretical mechanics in Galilei's and Newton's works.

Following thousands of years of observation and experiment, the knowledge was summed up to yield the concepts of time and space as entities which were independent both mutually and of any other phenomenon in the universe, of any state or transformation. The absolute space and the universal time, whose lapse is uniform at any point of the universe, related to nothing, were influenced by nothing. The absolute space and universal time exist independently.

A unique variable of the whole universe, the pre-relativistic time is so conceived that two events are simultaneous, in a certain sequence or separated by a certain interval regardless of observer, his position, or his motion. There is everywhere the same past, present and future.

The general progress and main advances in experimental techniques have determined the coming into being of the theory of relativity; its first effect was to revolutionize the concepts of time and space, with obvious consequences in all fields of knowledge.

The space-time interdependence came next; it is elegantly expressed by Minkowski's space, whose metric is established by the well known equation:

$$ds^2 = c^2dt^2 - (dx^2 + dy^2 + dz^2), \qquad (B.1)$$

where, according to Part A of this book, ds is the space-time interval, $x = u_1$, $y = u_2$, $z = u_3$ are the space coordinates and $ict = u_4$ the time coordinate ($i = \sqrt{-1}$, c is the speed of light, and t is the time). This offers the possibility of expressing the invariant ds by the equation:

$$ds^2 = -(du_1^2 + du_2^2 + du_3^2 + du_4^2), \qquad (B.2)$$

which suggests an Euclidean metric (the Minkowski space is pseudo-Euclidian) and the much discussed spatialization of time.

Indeed, time and space are not only indestructibly related but, moreover, time is regarded as a fourth dimension of space which thus becomes four-dimensional. The simultaneity and the order of succession of some events depends on the observer, on his state of motion.

The interdependence of space and time is achieved through motion without which time could not be conceived. The concept of time may be reached only by means of transformation.

The set of transformations includes as a subset the ensemble of mechanical motions.

However, as noted in Part A, the tendency of total identification of time with space (more precisely with a fourth dimension of space) is unacceptable in a truly physical sense.

It is true that modern physics, unlike classical pre-relativistic science, cannot conceive space without matter (which may assume one of the two basic forms: substance and energy field); matter is inconceivable without motion and motion leads directly to the concept of time. Thus space and time cannot be split. However their compulsory association, their intrinsic relationship, their permanent interdependence does not mean their identification.

The tendency to regard time exclusively as a fourth dimension of space , i.e., to spatialize time, may be opposed, with the following arguments, that hold true in the theory of relativity (both special and general):

(a) The irreversibility of time imposes a variation in one unique sense of the time coordinate. Unlike time, no space coordinate has an imposed sense of variation. There is return in space, but there is no return in time.

(b) The time lapse , i.e., the variation of the time coordinate of one phenomenon may not be stopped, no matter what the frame employed, while a zero variation of space coordinates may be achieved by a suitable choice of the frame that the motion is referred to (mechanical motion if space coordinates are involved).

In eqn. (B.1) the qualitative difference between time and space is reflected in the different sign of variation of the space coordinates and the time one, while in eqn. (B.2) the same qualitative difference is emphasized by the imaginary unit which is a coefficient for the time coordinate only.

On the other hand, statistical physics and quantum mechanics, which are based on a statistical determinism, i.e., accept states in probabilistic succession, assign a special structure with random character to the time of the microcosmos.

Finally, it is worth recalling that the theory of information recommends as criterion of defining, perceiving and expressing time the succession of information stored by a brain, or by any cybernetic system. In other words, the storage of information whose amount (and quality) varies continuously is taken as the basic transformation in defining time (the concepts of present, past and future included).

However, this procedure runs a risk: it leaves room for subjectivism.

Obviously these brief considerations regarding the time of classical (pre-relativistic) physics, the time of the theory of relativity, the concept of time as based on the phenomena in the microcosmos (studied by quantum mechanics), and that related to the theory of information achieve a presentation of the core of some viewpoints (considered here as worth noting) developed in the specialized literature.

In the following, besides the author's original contributions, the book considers to the extent deemed necessary, the previous approaches to the concept of time and attempts (even if not specified) to incorporate them.

As one of the main pillars of the contemporary scientific thinking, the theory of relativity plays a major role in any profound investigation of the concept of time.

Therefore, the theory of relativity was reviewed in the first part of this book. (The review was restricted mainly to relativistic mechanics, leaving aside relativistic electrodynamics, thermodynamics and quantum mechanics, because it was considered, on the one hand, that it suits our aims and, on the other, that mechanical motion is the simplest and most suggestive form of motion which involves the two entities: space and time). Moreover the way some motions were introduced and interpreted (see, e.g., section A.II.1.15) anticipates the approach of some objectives of this book.

B.I.2. Time as measure and consequence of motion

Whenever in a field of science a physical or mathematical model, that approximates the evolution of a class of phenomena or defines concepts of general and fundamental implications in the study of nature, ceases to be suitable, either due to the rough approximation it achieves or due to some new discoveries that contradict it, the model has to be either replaced or improved.

However, the modification of a general or particular model or the introduction of new concepts on some fundamental notions (which obviously exerts a major influence upon the understanding and study of the universe) may also be achieved without the existence of some stringent need when the conditions of general progress of knowledge and scientific thinking ensure the conditions for a synthesis. This brings about consequences that contribute to approaching the truth, which the human mind achieves, asymptotically through a long chaim of successive approximations.

Let us now present a new point of view on the concept of <u>time</u>. We start from the primary element: the phenomenon observed in the surrounding Nature.

B.I.2.1. <u>Motion</u>. The state of a material system Γ is defined by the values of the ensemble of physical quantities which characterize the state of the system in question.

Assume that an observer O measures these state quantities and finds $q_1, q_2, \ldots, q_\gamma$, where, by hypnothesis, γ is the number of physical quantities involved in the phenomenon.

The ensemble of values $q_1, q_2, \ldots, q_\gamma$ represents the <u>state</u> of the system they characterize. Two states differ from each other at least by one of the state quantities.

If a series of p measurements comes out with the values:

$$q_{1j}, q_{2j}, \ldots, q_{\gamma j} \quad (j = 1, 2, \ldots, p)$$

for the state quantities; if <u>at least</u> for one of them, e.g., q_i $(i \in \{1, 2, \ldots, \gamma\})$, one finds that, for $j \neq k$, $q_{ij} \neq q_{ik}$, the material system is in motion or, to say it more generally, is subjected to a transformation, i.e., it is not in an unique state. The term <u>motion</u> is given the general meaning of <u>transformation</u>. The mechanical motion is only one of the possible cases, i.e., that case when q_i $(i = 1, 2, \ldots, \gamma)$ are space coordinates. Obviously, in general they are generalized coordinates and in particular (for material points) they may be Cartesian, cylindrical or spherical coordinates.

It turns out that motion is expressed by ensembles of q_i values which define the set of states forming a succession of states. However, this is a point of some subtlety.

To define a state of the material system in question, the observer has to establish those values of q which correspond mutually. Thus the notion of concomitance and simultaneity are outlined.

To express this in detail, studying the system in question the observer O records a number γ of sets Q_i of values q_i ($q_i \in Q_i$; $i = 1, 2, \ldots, \gamma$) and notes a correspondence of these sets; this in the sense that, in fact, a set A of elements a_j ($j = 1, 2, \ldots, p$) is noted (provided that p measurements are made), each element a_j being an ensemble of values q_{ij} ($i = 1, 2, \ldots, \gamma$), i.e., representing a state of the system Γ. One can say that for the observer O these values q_{ij} (that form an element $a_j \in A$) are simultaneous.

One can notice that there could be no certainty about the following facts: (a) various observers note the same values q_{ij} to make up an element $a_j \in A$, i.e., for various observers the same values q_{ij} of the state variables are simultaneous; (b) even for one and the same observer O, the simultaneity of quantities q_{ij} does not depend on the way measurements were effected.

Thus, on the one hand, it is not certain at all that the states of the material system are the same for various observers and, on the other hand, it is uncertain that one and the same observer notes states that are independent of the way the measurements were performed, of their site and of the state of the observer.

It should also be remarked that the sets Q_i may possibly be of the power of the continuum, i.e., the q_i can vary continuously (which is very often the case). Then, a_j has obviously a continuous variation.
Because the elements $a_j = \{q_{ij}, q_{2j}, \ldots, q_{\gamma j}\}$ define a state of the system Γ (noted by the observer O), the set A ($a_j \in A$), which has to be ordered, as is shown below, represents in fact the succession of states, i.e, the motion of Γ. Meanwhile, $a_j = \{q_{ij}\}$ and $a_h = \{q_{ik}\}$ define various states if for at least one $i \in \{1, 2, \ldots, \gamma\}$, $q_{ij} \neq q_{ik}$.

The notion of succession of states, implicitly of motion of the material system in question, has to be rigorously correlated with the ordering of the set A.

These general statements are basically true whether the sets Q_i of values of state variables q_i ($q_i \in Q_i$) are continuous or discrete. If a set Q_i contains more than one element q_i, the quantity q_i varies.

From all these statements the conclusion emerges that an observer O establishes a number c of correspondences between the elements q_i of sets Q_i in terms of the states characterized by the ensemble $a_j = \{q_{1j}, q_{2j}, \ldots, q_{\gamma i}\}$ of values of the state quantities q_{ij} ($i = 1, 2, \ldots, \gamma$). An elementary reasoning leads to the conclusion that the largest c (in case of γ state quantities) is:

$$c = \sum_{m=2}^{\gamma} C_{\gamma}^{m} = \sum_{m=2}^{\gamma} \frac{\gamma(\gamma-1)\ \ldots\ (\gamma-m+1)}{m!}. \quad (B.3)$$

It may be assumed that these correspondences are expressed by functions like:

$$f(q_{i1}, q_{i2}) = 0, \ f(q_{i1}, q_{i2}, q_{i3}) = 0, \ \ldots, \ f(q_{i1}, q_{i2}, \ldots, q_{i\gamma}) = 0, \tag{B.4}$$

in which obviously $i_k \neq i_s$ and $i_k \in \{1, 2, \ldots, \gamma\}$ and $i_s \in \{1, s, \ldots, \gamma\}$.

Implicit forms were employed in eqn. (B.4); but using explicit ones is also possible.

It is worth remarking that functions (B.4) may contain state quantities whose variations are related by causality and state quantities whose variations are not related by causality; in the latter case the correspondence between their values is established only by the succession of states noted by the observer and implicitly by the values that the q_i's take in the ensembles $a = \{q_1, q_2, \ldots, q_\gamma\}$ which characterize these states. However, between q_i and q_k there may or may not exist causality relations, i.e, the variation of one of them does or does not condition the variation of the other one. The statement holds true for 3, 4, ..., γ state quantities.

It should be mentioned again that nothing ensures that:

(a) two different observers note the same correspondences between the state variables q_i;

(b) even for one and the same observer, the functions (B.4) are independent of the state of the observer.

The domains of definition and the co-domains of functions (B.4) may be continuous or discrete sets, according as q_i $(i = 1, 2, \ldots, \gamma)$ take continuous or discrete values or as Q_i $(q_i \in Q_i)$ are continuous or discrete sets.

According to the foregoing, if a set Q_i' contains only <u>one</u> value q_{oi} of the state quantity q_i, this means that this latter remains <u>constant</u> during the motion studied.

It is worth noting that an observer O does not generally limit his observation to the motion of one single material system. Observation may extend to n systems Γ_s (s = 1, 2, ..., n). Practically, n is always finite (even though sometimes very large) but in principle, for generalization, n may be assumed infinite.

Consider $q_{s1}, q_{s2}, \ldots, q_{s\gamma_s}$, the state quantities of the material system Γ_s.

The number of these quantities is γ_s; and in general it is possible that $\gamma_s \neq \gamma_k$ and $\gamma_s = \gamma_k$ (with $s \neq k$, where γ_k refers to system Γ_k).

For a material system Γ_s, the number γ_s represents the number of <u>degrees of freedom</u> if q varies independently, that is, unrelated to causality factors, in relations of type (B.4). In case of a number of such material systems Γ_s (s = 1, 2, ..., n), the state quantities q_s, whatever s and i, may be quantitatively identical or different (i = 1, 2, ..., γ_s).

All previous considerations regarding the material system Γ and its state quantities q_i hold true for all material systems Γ_s (s = 1, 2, ..., n) and their state quantities q_{si} (i = 1, 2, ..., γ_s). In what follows, some of these considerations are developed and some new ones are introduced.

Consider a state quantity q_{si} which is involved in expressing the states of a material system Γ_s and consider the set of all q_{si} values, Q_{si}.

It is said that quantity q_{si} <u>varies continuously</u>

and, implicitly, that Q_{si} is a <u>continuous set</u>, if for any arbitrary positive number ε_i, however small, two observations can be made to measure values q'_{si} and q''_{si} such that:

$$|q'_{si} - q''_{si}| < \varepsilon_s, \tag{B.5}$$

<u>no matter</u> what $q'_{si} \in Q_{si}$ (or, which is the same, no matter what $q''_{si} \in Q_{si}$).

If however the observations are made and measurements taken and, whatever the values $q_{sij} \in Q_{si}$ and $q_{sik} \in \theta_{si}$, one finds out that:

$$|q_{sij} - q_{sik}| = \lambda_{jk} \nu_{si}, \tag{B.6}$$

where $\lambda_{jk} = 0$ for $j = k$ and $\lambda_{jk} \in N$ (the set of natural numbers) for $j \neq k$ and ν_{si} does not depend on j and k (i.e., ν_{si} is a constant quantity for a given q_{si}, that is, given Q_i), then the state quantity q_{si} undergoes <u>quantized</u> variation and the <u>quantum</u> of this variation is ν_{si}.

The immediate consequence is that any value of the state quantity q_{si} is an integer multiple of ν_{si}. Thus in such a case:

$$q_{si} = \lambda \nu_{si}, \tag{B.7}$$

with $\lambda \in N$.

Obviously, for any interval,

$$(\eta, \rho) \subset (k\nu_{si}, (k + 1) \nu_{si}) \tag{B.8}$$

one can write immediately:

$$q_{si} \notin (\eta, \rho), \tag{B.9}$$

for k natural or zero.

However, there may exist another state quantity which also refers to Γ_s ($q_{s\alpha}$, $\alpha \in \{1, 2, \ldots, \gamma_s\}$ but $\alpha \neq i$) or to Γ_r, another material system observed by O ($q_{r\beta}$; $\beta \in \{1, 2, \ldots, \gamma_r\}$), and, although they may have a quantized variation, there exists a natural number λ or λ' such that:

$$q_{s\alpha} = \lambda\nu_{s\alpha} \in (\eta, \rho),$$
$$q_{r\beta} = \lambda'\nu_{r\beta} \in (\eta, \rho), \tag{B.10}$$

because $\nu_{s\alpha} \neq \nu_{si}$ and $\nu_{r\beta} \neq \nu_{si}$, except for the cases of coincidence, $\nu_{s\alpha}$ and $\nu_{r\beta}$ being obviously the quanta of variation of $q_{s\alpha}$ and $q_{s\beta}$, respectively, in whose expressions appropriate units are employed.

In case any state quantity varies continuously, there is an infinity of its values $q \in (\eta, \rho)$, whatever the interval (η, ρ).

Finally, it is also possible that some state quantities q_{si} exhibit random variations on a discrete set Q_{si}.

Q_{si} is obviously a discrete set even if q_{is} exhibits quantized variation, according to eqn. (B.7). Evidently Q_{is} is a continuous set only if q_{is} varies continuously.

It may also be affirmed that the totality of material systems Γ_s ($s = 1, 2, \ldots, n$) observed by O represent the universe controlled by observer O. Within this universe, the observer notes the states and motions of material systems; the motions assume variations of the state quantities. These variations may be continuous or discrete.

If all state quantities, q_{is}, of a material system Γ_s vary continuously, the motion of Γ_s is continuous.

If at least one of the state quantities, q_s, of a material system Γ_s exhibit a discrete variation, the system is in discontinuous motion.

B.I.2.2. The clock. We have already talked about the variation of state quantities of a system Γ_s. The variation of a state quantity, q_{is}, means that the set Q_{is} of its values includes more than one number. Q_{is} may be a finite or an infinite set; in the latter case it may be discrete or continuous.

The variation of state quantities means in fact the motion of the system they refer to.

The problem now is to find a measure of motion, a possibility of expressing quantitatively a phenomenon referred to as motion. As specified, the term "motion" refers to any transformation be it a rigorous mathematical representation (belonging to higher stages of the process of knowing) or an approximate estimation required by an incipient ordering and systematization of notions resulting from conscious interaction with the universe (at a primary or superficial stage of knowledge).

The value of any physical state quantity and any variation of it may be expressed only on the basis of units, i.e., by comparison with a certain physical quantity of the same entity as that measured, a physical quantity whose value remains obviously constant (and is chosen as unit).

Moreover, any representation of a phenomenon

requires comparison, the existence of at least one reference (in the widest sense of the word) and of an ensemble (system) of units.

Any reference assumes the existence of an origin chosen arbitrarily or one with absolute character. In case of mechanical motion, there may not be an absolute reference. The references regarded as fixed, with respect to which motion is called (improperly) absolute, are the consequence of an agreement.

They are, in fact, basic references to which motions studied are referred and, in case of another choice of the reference frame, may become mobile. The absolute zero could be considered as the absolute reference for temperature variation.

To express a state S of a material system Γ_s it should be referred to another state S_0 which acts as reference.

The state S is defined (with respect to state S_0 by variations Δq_{si} of the state quantities q_{si} between S_0 and S.

Obviously the quantitative expression of variations Δq_{si} $(i = 1, 2, \ldots, \gamma_s)$ implies the utilization of units.

And now we come to the main point of the presentation.

The ordered set of all states of a material system is the succession of states, i.e., the motion of the material system in question.

The central problem which awaits elucidation is the analysis of the way the states are ordered.

The observer O achieves the ordering of the states of a material system by means of a comparison process as well. That is, he notes (and measures) the motion of the system. The observer has to establish a correspondence between the set θ of values of a state quantity τ (that is $\tau \in \theta$) of a material system C and the sets Q_{si} of values q_{si} ($q_{si} \in Q_{si}$) of the state quantities of the system Γ_s whose motion is studied, i.e., has to compare the motion studied with another motion.

The totality of values q_{si} the observer O sets in correspondence with the <u>same</u> value of τ form a state of Γ_s and these values are <u>simultaneous</u>. Thus an ensemble of values of state variables q_{s1j}, 1_{s2j}, ..., $q_{s\gamma_s j}$ correspond to the <u>same</u> values of τ_j and make up an element $a_{sj} = \{q_{s1j}, q_{s2j}, \ldots, q_{s\gamma_s j}\}$ of the set A_s ($a_s \in A_s$); and the set A_s expresses the set of all states of the Γ_s system.

The desired ordering for the set of states of Γ_s, implicitly for sets Q_{si} and A_s, is achieved if Γ_s satisfies the following conditions:

(1) For all material systems Γ_s observed by O, i.e., for the whole of observer O's universe, there do not exist two (or more) values of the state quantity q_{si} (whatever it is) which correspond to the same value of τ.

The consequence of this statement is that, in the case when at least one of the state quantities of one of the material systems observed takes values which form a continuous set, θ ($\tau \in \theta$) has to be a continuous set. If all sets Q_{si} of values of state quantities q_{si} (whatever the system Γ_s observed) are discrete sets, θ could (in principle) be a discrete set such that, taking any two values q_{sij} and q_{sik} of a state quantity q_{si} (no matter which state quantity and material system it refers to), if τ_j is the value of τ corresponding to q_{sij} and τ_k the value corresponding to q_{sik}, then $q_{sij} \neq q_{sik}$ implies $\tau_j \neq \tau_k$. Assuming that all state quantities are quantized, this amounts to saying that, for $q_{si} = \lambda\nu_{si}$, where λ is a natural number and ν_{si} the quantum, for any λ, s and i, the following relation holds:

$$\tau_{\lambda + 1} - \tau_{\lambda} = \alpha\nu, \tag{B.11}$$

where $\tau_{\lambda + 1}$ corresponds to $q_{si} = (\lambda + 1)\nu_{si}$ and τ_λ to $q_{si} = \lambda\nu_{si}$; ν are increments of τ and α is a natural number (whose minimum value is 1).

It is important to notice that if eqn. (B.11) is obeyed for a number of state quantities $\gamma' = \sum_{s=1}^{n} \gamma_s$, it is easy to show that, whatever the interval of real numbers (a, b) considered with $|b - a| < \varepsilon$, the probability that $\alpha\nu \in (a, b)$ (for any natural number α) increases with γ' and decreases with ε. No matter how small ε is, the probability that $\alpha\nu \in (a, b)$ tends to one if γ' tends to infinity. Obviously this is so because one should observe the condition that to each state of the observer's universe corresponds a value τ and only one.

The outstandingly important conclusion one can draw is the following: if the observer studies an infinite number of state quantities q_{si} (which may be of an infinite number of qualities, e.g., spaces, temperatures, pressures etc.), θ has to be a continuous set even in case each state quantity takes values in a discrete set. Most often, the set θ is considered continuous. Thus it may offer a distinct τ value, whatever the sets Q_{is} ($q_{is} \in Q_{is}$), for any state of the universe (one phase differs from another by at least one of the values of q_{is}).

(2) The set θ is ordered. This amounts to stating that, for a series of events $E_1, E_2, \ldots, E_p$ related by causality factors, i.e., E_β is caused by $E_{\beta - 1}$ and causes $E_{\beta + 1}$ (these events being, e.g., states of the same system or different systems), if τ_β corresponds to E_β and τ_α to E_α, $\beta > \alpha$ implies $\tau_\alpha < \tau_\beta$. Obviously $\tau_\alpha \in \theta$, $\tau_\beta \in \theta$.

It means that any variation $\Delta\tau$ of τ noted by O is positive ($\Delta\tau > 0$).

In these conditions, C is referred to as a clock and the state quantity whose values τ form the set θ is referred to as timing quantity or timing parameter.

Association of the two conditions which have to be fulfilled by a system in order to be a clock, i.e., to be useful to an observer in ordering the states of the other system, that is to measure and express motion, shows clearly that, as already mentioned, if at least one set Q_{is} ($q_{is} \in Q_{is}$) is continuous, obviously θ ($\tau \in \theta$) has to be continuous. However, if all sets Q_{is} are discrete, let $q_{is} = a_{is}$ and $q_{is} = b_{is}$ be any two values of the quantity q_{is} and τ_a and τ_b the corresponding values of the timing parameter. It is assumed that $a_{is} < b_{is}$ and $\tau_a < \tau_b$, so that $\tau_b - \tau_a = \Delta\tau > 0$. It is also assumed that, because all sets Q_{is} are discrete, θ could be a discrete set, the values of τ being quantized. Any $\tau = \alpha\nu$, with ν a quantum. Consider now n_{ab}, the number of all values q_{is} (for any i and s) which correspond to values $\tau_{is} \in (\tau_a, \tau_b)$. Obviously, the τ_{is} are in an overall number of n_{ab} and $\Delta\tau \geqslant n_{ab}\nu$. However (according to a simple probability calculation), the higher the number γ' of all state quantities, the higher n_{ab}. If γ' tends to infinity, n_{ab} tends also to infinity.

As $\Delta\tau$ may always be kept finite, it follows that, for an infinite number of quantities of the systems observed (i.e., an infinite number of systems observed), the quantum ν has to tend to zero, that is, τ varies continuously (θ is a continuous set).

The state of the clock C which corresponds to $\tau = 0$ may generally be chosen arbitrarily. It is called the origin of the clock. The motion of C is a standard motion referred to as clocking (timing) motion.

The way the observer achieves the ordering of the set of states of the system, establishes their sequence, i.e., forms the image of the motion of systems, is now obvious. The observer compares any motion under study with the clocking motion. However, for a rigorous interpretation and presentation of this aspect one should notice that eqns. (B.4) may be extended throughout the whole universe observed by O. That is, eqns. (B.4) should not necessarily relate only the state quantities of the same system but of different systems as well. The total number C_n of such functions is, in this case, given by a relation which is similar to (B.3):

$$c_n = \sum_{m=2}^{\gamma'} C^m_{\gamma'}, \quad \gamma' = \sum_{s=1}^{n} \gamma_s, \qquad \text{(B.12)}$$

and the functions relate between them 2, 3, ..., γ' variables which represent the values of all state quantities of all systems (n in all) observed by O, in other words, all the state quantities in the universe of O. One of the systems in the universe of O is the clock C and one of the γ' sets of values whose (B.4)-type functional dependence is established by O is θ, the set of values of the clocking parameter τ.

From among the functions recalled here, only those are retained which establish the relation (correspondence) between τ and one of the state quantities q_{is} ($i = 1, 2, \ldots, \gamma_s$) of the system Γ_s ($s = 1, 2, \ldots, n$), that is all state quantities in the universe of O.

These functions are:

$$f_{is}\ (\tau,\ q_{is}) = 0; \qquad i = 1, 2, \ldots, \gamma_s, \tag{B.13}$$

which can be solved as

$$q_{is} = \phi_{is}\ (\tau); \qquad s = 1, 2, \ldots, n. \tag{B.13'}$$

Thus, there are a total of $\gamma' = \sum_{s=1}^{n} \gamma_s$ such functions; and obviously γ' may be finite or infinite (depending on θ's universe).

There is an univocal or biunivocal correspondence between sets θ and Q_{is} ($\tau \in \theta$, $q_{is} \in Q_{is}$). One single q_{is} value corresponds to each τ value, given the conditions τ has to fulfill in order to be a clocking parameter (conditions enumerated above).

Thus, all state quantities are uniform functions of the clocking parameter. This dependence of state quantities on τ amounts to a dependence on τ of all states. The dependence is formal, strictly quantitative - mathematical, not necessarily causal. In particular, some ϕ_{is} functions may be constant.

A state of a system Γ_s is expressed by an ensemble

$a_s = \{q_{1s}(\tau), q_{2s}(\tau) \ldots q_{\gamma_s s}(\tau)\}$ of values of q_{is} quantities corresponding to the same value τ.

All values q_{is}, regardless of i and s, which correspond to the same τ are simultaneous for the observer O. Thus the state is determined by the simultaneous state quantities. Consider τ_p, a value of the clocking parameter τ. If $\tau_t < \tau_p$ ($\tau_t \in \theta$), a state, S_t corresponding to τ_t determined by an ensemble of state quantities $a_{st} = \{q_{1s}(\tau_t), q_{2s}(\tau_t) \ldots q_{\gamma_s s}(\tau_t)\}$ is anterior to the state which is characterized by an ensemble of state quantities $a_{sp} = \{q_{1s}(\tau_p), q_{2s}(\tau_p) \ldots q_{\gamma_s s}(\tau_p)\}$ of q_{is} quantities. The totality of such states S form the past of the state S_p determined by a_{sp}.

Similarly, a state S_v given by the ensemble a_{sv} of state quantities of the system in question, $a_{sv} = \{q_{1s}(\tau_v), q_{2s}(\tau_v) \ldots q_{\gamma_s s}(\tau_v)\}$, i.e., corresponding to a value $\tau_v > \tau_p$, is subsequent to the state a_{sp} and all such states S_v form the future of state S_p.

The states simultaneous with S_p represent the present of S_p.

One should not forget that in these definitions the ordering of the set θ was taken into account. Any value τ has then a distinct significance for the observer O. It corresponds to one state of the whole universe controlled by O.

Note. The simultaneity of states, the succession, the manner of ordering as a function of the clocking parameter τ, i.e., the motion in general as well as its past and future with respect to a certain state, are the result of the comparison (made by the observer) between the motion studied and a standard motion referred to as clocking motion, i.e., the motion of clock

C, implying the set of values of the timing parameter τ. Thus, one reaches the functions (B.13') which contain the complete image about the motion of the universe the observer O controls or the image of a certain system within this universe.

However, regarding the correspondence between the values of the state quantities of various systems and the timing parameter τ, i.e., in fact, the functions (B.13') or else the motion noted by the observer, one should consider carefully the simultaneous or successive character of some states. Even their belonging to the past, future or present (if the observation is referred to a certain given state) has to depend on the way the observer gets the information about the systems studied. Thus the findings of an observer O are functions of the relations between the observer and the system being observed and of the agent which carries information to the observer (if such an agent exists).

It turns out that, in general, the findings of two (or more) observers could differ from each other. Thus, one comes to the relative character of motion and of those entities that the study of motion has led to.

Such an approach and (mostly) its subsequent development will make from the relativistic (Einsteinian) model of nature something as natural as possible.

Obviously, the observer O may note the existence of more than one material system C which fulfill all requirements to be employed as a clock, i.e., of a set K of such clocks C_u ($u = 1, 2, ..., \mu$) whose motions may be considered clocking motions. The set K may be finite or infinite.

Let τ_u be the clocking parameter of clock C_u and θ_u the set of its values ($\tau_u \in \theta_u$). Obviously, this implies that θ_u (i.e., the values of τ) observes the conditions required by the standard motion referred to as clocking motion, whatever u, i.e., whatever the element of set K.

Because, in fact, the τ_u are state quantities, there is a number $C_\mu^2 = \frac{\mu(\mu - 1)}{2}$ of functions:

$$\tau_w = \phi_w(\tau_u), \tag{B.14}$$

which observer O establishes and which are nothing but (B.13')-type functions.

In eqn. (B.14), $u \in \{1, 2, \ldots, \mu\}$ and $w \in \{1, 2, \ldots, \mu\}$, μ being the total number (finite or infinite) of clocks which come under O's observation.

The fulfillment of the conditions known to be imposed upon some clocking parameters ensures that functions (B.14) establish a biunivocal correspondence between θ_u and θ_w ($\tau_u \in \theta_u$; $\tau_w \in \theta_w$).

The set of clocks $K = \{C_1, C_2, \ldots, C_u\}$ may be divided into two equivalence classes which are subsets formed of clocks mutually equivalent.

All clocks for which an observer O establishes a linear functional dependence like:

$$\tau_w = a_{u_w}\tau_u + b_{u_w}, \tag{B.15}$$

where a_{u_w} and b_{u_w} are constant, are referred to as equivalent with respect to the observer O.

Obviously (B.15) is a particular case of (B.14).

<u>Conclusions</u>:

If the units employed to measure τ_u and τ_w are u_o and w_o respectively, then a is given in units like $\frac{w_o}{u_o}$ and b in w_o.

If L_α and L_β are two subsets of K forming two equivalence classes, then $L_\alpha \cap L_\beta = \emptyset$ (the empty set).

If the total number of equivalence classes in K is m, then $\bigcup_{\alpha=1}^{m} L_\alpha = K$.

If $C_u \in L_\alpha$ then necessarily $C_u \notin L_\beta$ for $\alpha \neq \beta$.

B.I.2.3. Examples of mechanical motion. Some examples are supplied next, related to the motions introduced above to make the presentation clearer, although it is still in an early stage.

These examples refer to mechanical motion, firstly, because this is a book of mechanics (implications in the other fields of research of nature and epistemologic consequences are also dealt with) and, secondly, because mechanical motion is the most intuitive form of motion in the universe.

In the case of mechanical motion the state quantities are the coordinates that define the positions of material systems with respect to the reference frame chosen. They may be Lagrangian, Cartesian, cylindrical or spherical coordinates. The sets of values are continuous.

The clock chosen may be Earth itself, the clocking motion being in this case the revolution of the planet about the Sun or its rotation around its own axis. Likewise, artificial clocks may be constructed, material systems whose motion is imposed, in which the variation of one quantity (e.g., an angle) achieves an equivalence with the clock employing the motion of the planet or of any star. These are usual clocks (the motion is the rotation of some hands about the fixed point on a dial).

Likewise, in order to prefigure a matter of major importance, let us notice that the propagation of an electromagnetic wave (light) may be ascribed a clocking character, i.e., it may be a standard motion which may be helpful in ordering the observations.

Some clocking motions are cyclic, i.e., τ varies periodically (the yearly and daily motion of the planet, the motion of the clock hands etc.). This does not conflict with the ordering of the set of values of the clocking quantity, because, even if the value of the clocking parameter is reproduced after each period, the value associated with the number n' of cycles, considered from a chosen origin to the cycle when the given value τ is recorded, gets a unique character. In other words, even if a clocking motion assumes several (perhaps an infinity of) identical τ values, it possesses one single pair of n' and τ values. E.g., any event may be set in univocal correspondence with a position of our planet (defined against the Sun), however it is necessary to evaluate the number of rotations the Earth has completed around the central star (counting of rotations is done from a certain origin of cycles and is measured in years).

Often one can find, even in these cases - as the clocks with cyclic (periodical) motions are - some way of timing which does not resort to the periodical character.

To remain a while longer with the example of the Earth (with its motion) employed as a clock, one can find (although it would be tiresome) the correspondence between a certain event and the surface area swept by the focal radius, with the origin at the Sun, defining the position of Earth from a certain origin. Or else one can resort to angles of rotation whose values may be of the order of $n\pi$ radians, though with n high (we refer here to the rotation of the planet about its own axis).

B.I.2.4. Time as an abstraction of general motion. Utilization of any clocking parameter τ implies such drawbacks as, e.g.:

(a) The qualitative aspect of τ and the units employed in expressing it have always to be specified.

(b) If several clocks with various state variables τ are employed, this has to be specified, and so has the relation between them, even if they belong to the same equivalence class.

(c) It is difficult to find two (or more) clocks which belong mathematically to the same equivalence class, i.e., whose timing quantities obey the equivalence condition (B.15).

However, this inconvenience may be removed by employing just one clock, but this is not always possible and it is anyway uncomfortable.

(d) Utilization of one physical state quantity τ as clocking parameter is logically possible in any kind of ordering of some processes in nature, of some successions of states of some motions, only if the observer sets τ in correspondence with the state quantities of the material systems that form the observer's universe. Thus the clock has to belong necessarily to the observer's universe.

Thus, it is difficult or impossible to express simultaneities and successions of some motions from the past anterior to the smallest value τ they may refer to (it is worth recalling that θ is increasingly ordered), or the future following after the largest τ which is acceptable from a physical point of view. This is obviously so in order not to disturb the logic in developing our reasoning.

E.g., if we define one year as a variation of 2π radians of the angle of the Sun-Earth focal radius with a fixed direction that passes through the center of Sun and is contained in the plane of the terrestrial orbit (a complete revolution of the planet), it is not possible without sacrifice of rigour to employ the clock Earth (i.e., the clocking motion of Earth about the Sun) to achieve ordering (simultaneity, succession etc.) of some motion occurring of the Earth.

All these drawbacks of the exclusive utilization of clocking parameters supplied by various physical state quantities - physical parameters that are utilized in ordering of states and giving the image of motion - as well as the natural need of achieving a synthesis and generalization of the motion of clocking variable or parameter, have led to the notion of time.

Thus, by a general synthesis of all clocks belonging to an equivalence class, one reaches an abstract variable, an abstract clocking parameter (not related to any clock) called time.

The set of its values is continuous and each of its values is called a moment. The usual notation for time is, as known, t.

Let L be the set of all clocks belonging to some equivalence class.

Now a very important analysis is needed.

It was already underlined that the correspondence

between the values of state quantities, i.e, the states of the universe and their succession, the way of studying the general motion and with it the clocking motions, all depend on the state relationship between the phenomena observed and the observer.

This is due to the way the observer receives information about the systems under study, via at least one information carrier. Obviously, the state, the succession of states, i.e, the motion of a certain system, could be noted only by the influence that the system exerts upon the observer, that is, by altering this influence. However, if the system observed has no direct contact with the observer, the influence, i.e., the information received, has to be carried by a physical agent. This information carrying agent is most often the electromagnetic wave, light; in the following reasoning it will be considered as the only information carrier for reasons easy to understand which will be analyzed in the chapter dealing with velocity.

This means that the establishing of any quantity of any motion depends on the observer's motion, position and variation of position correlated with the motion of the information carrier; therefore any equivalence class L_α of clocks depends on the observer.

The property belonging to a set L_α of equivalent clocks holds true only for a certain (or certain) observer(s) not for any observer.

All the above considerations have evidently referred to the mechanical motion of the observer and to the mechanical motion (propagation) of the information carrier. We have the feeling that these considerations are nothing but one of the main aspects of the theory of relativity. The relativistic character of the universe is in fact the relativistic character of observations, measurements, experimental findings.

Returning now to the set L_α of clocks C_u which form an equivalence class, their parameters τ_u ($u = 1, 2, \ldots, \nu_\alpha$) satisfy equations of the type (B.15), i.e.:

$$\tau_{i\alpha} = a_{ij\alpha}\,\tau_{j\alpha} + b_{ij\alpha}, \quad \begin{matrix} i = 1, 2, \ldots, \nu_\alpha \\ j = 1, 2, \ldots, \nu_\alpha \end{matrix} \qquad \text{(B.16)}$$

where a_{ij} and b_{ij} are constants, if L_α contains, according to the observer O, a finite or infinite number ν_α of clocks. Obviously, if, in particular, $i = j$, then $a_{ij\alpha} = 1$ (and dimensionless) and $b_{ij\alpha} = 0$.

One should not forget that this linear relationship is in general approximate.

It is difficult to find two perfectly equivalent clocks. However, when the deviations are small, one accepts the ideal situation in which eqn. (B.16) is satisfied.

The abstract variable t - the abstract clocking parameter called time - is attached to a set of clocks which form an equivalence class, e.g., the L_α set. Thus, if $C_{u\alpha}$ is a clock of parameter $\tau_{u\alpha}$ and if $C_{u\alpha} \in L$, then:

$$\tau_{u\alpha} = v_{u\alpha}\, t_\alpha + k_{u\alpha}; \quad u = 1, 2, \ldots, \nu_\alpha; \qquad \text{(B.17)}$$

where $v_{u\alpha}$ and $k_{u\alpha}$ are constants and t_α is time. The subscript α next to t emphasizes that the abstract variable named time has resulted from the synthesis of the comparison of motions in nature with those of clocks in the L_α equivalence class. To any value of time t_α, i.e., to any <u>moment</u> corresponds a value (and only one) of the clocking parameter $\tau_{u\alpha}$, implicitly a state (and only one) for any system observed by O, that is, a unique state of the universe controlled by the observer.

One can always choose the sense of the measurement of $\tau_{u\alpha}$ such that the constant $v_{u\alpha}$ is positive ($v_{u\alpha} > 0$ for $u = 1, 2, \ldots, \nu_\alpha$). For the present phase of the presentation only <u>non-cyclic</u> clocking parameters are considered; the ones whose character is periodic are analyzed later.

It was seen that, with the notations in (B.16), the units employed in expressing $\tau_{i\alpha}$ and $\tau_{j\alpha}$ are $u_{i\alpha}$ and $u_{j\alpha}$, respectively; so the units for $a_{i\alpha}$ and $b_{i\alpha}$ are

$u_{i\alpha}/u_{j\alpha}$ and u_i.

The problem now arises as to the dimensions in eqn. (B.17), which amounts to the need of adopting some units for time. Because time has resulted from the generalizing synthesis of all clocks in the L_α equivalence class, to define the unit one has to resort to one of the clocks of class L_α.

Before anything, the notions of <u>time interval</u> and <u>duration</u> have to be defined first.

The measure of the set of all values of time between two moments is referred to as the <u>time interval</u>. If the two moments which delimit a time interval are exactly the first state (there is no state anterior to it according to any clock employed by the observer) and the last state (there is no other state following after it, a fact also established with any clock) of any one phenomenon (a motion, a phase of a motion), then the interval in question is the <u>duration</u> of the phenomenon. Thus any duration is a time interval, and any time interval associated with a phenomenon becomes the duration of that phenomenon.

A certain time interval, a certain duration, is taken as unit for measurements of time; the unit may as well be a fraction of a certain duration.

Therefore we resort to a certain phenomenon which is a timing motion. For this purpose one can employ any clock $C_{u\alpha} \in L_\alpha$.

According to (B.16) the variations of two equivalent timing parameters (the clocks belong to the same equivalence class L_α), $\tau_{i\alpha}$ and $\tau_{j\alpha}$, are $\Delta\tau_{i\alpha}$ and $\Delta\tau_{j\alpha}$ related by the equation:

$$\Delta\tau_{i\alpha} = a_{ij\alpha}\, \Delta\tau_{j\alpha}. \tag{B.18}$$

E.g., to close an orbital ellipse, the centre of mass of the planet in its revolution about the central star takes as long as 365.2422 rotations about Earth's own axis.

Taking now as timing parameters the two motions of the Earth: $\tau_{i\alpha}$ - the surface area swept by the focal radius; and $\tau_{j\alpha}$ - the angle of rotation of the planet about its own axis, these timing quantities are equivalent (to a certain approximation). Let A_o be the surface area swept by the polar radius during a complete rotation of the planet, i.e., the area of the ellipse which represents the orbit of the Earth.

Then eqn. (B.18) becomes:

$$\Delta\tau_{i\alpha} = \frac{A_i}{365.2422}\, \Delta\tau_{j\alpha} \qquad \text{(B.19}$$

and one can observe that:

$$a_{ij\alpha} = \frac{A_0}{365.2422} = \frac{A_0 \cdot 10^4}{3\ 652\ 422} \frac{[L^2]}{\text{rad}} \tag{B.20}$$

(expressed in surface units $[L^2]$ over radian). In fact, the radian being a length ratio (without dimension), $a_{ji\alpha}$ in (B.20) takes the dimensions of a surface.

Preserving the same timing variations, i.e., the variation of the surface area swept over by the polar radius from the Earth's centre (when the origin is at the Sun's centre) of parameter $\tau_{i\alpha}$ and the rotation of the planet about its own axis of parameter $\tau_{j\alpha}$, consider now eqn. (B.17). This, like any relation between physical quantities, assumes the existence of one unit (with its multiples and submultiples) for each physical quantity, and, at the same time, it may serve to define some of its units.

For this purpose, consider a certain time interval Δt_α. It is defined as corresponding to the variation $\Delta\tau_{u\alpha}$ to which it is related, according to (B.17), by the equation:

$$\Delta\tau_{u\alpha} = v_{u\alpha}\ \Delta t_\alpha; \qquad u = 1, 2, \ldots, \nu_\alpha, \tag{B.21}$$

which holds for any clock $C_{u\alpha} \in L_\alpha$.

Now consider two clocks $C_{i\alpha}$ and $C_{j\alpha}$ whose parameters are $\tau_{i\alpha}$ and $\tau_{j\alpha}$. For the same Δt_α, one gets:

$$\Delta\tau_{i\alpha} = v_{i\alpha}\ \Delta t_\alpha, \tag{B.21'}$$

$$\Delta\tau_{j\alpha} = v_{j\alpha}\ \Delta t_\alpha, \tag{B.21''}$$

that is:

$$\Delta\tau_{i\alpha} = \frac{v_{i\alpha}}{v_{j\alpha}} \Delta\tau_{j\alpha}. \quad \text{(B.22)}$$

Comparison of (B.18) and (B.22) yields:

$$a_{ij} = \frac{v_{i\alpha}}{v_{j\alpha}}. \quad \text{(B.23)}$$

And, if in particular $\tau_{i\alpha}$ is the area swept over by the Sun-Earth polar radius and $\tau_{j\alpha}$ the angle of rotation of the planet about its own axis, according to eqn. (B.20), one gets:

$$\frac{v_{i\alpha}}{v_{j\alpha}} = \frac{A_0 \cdot 10^4}{3\ 652\ 422}, \quad \text{(B.23')}$$

and the number which forms the right-hand side of this equation has the dimension of a surface area, the same units employed to express surface area A_0 of the ellipse which is the trace of the centre of mass of the planet in its orbital motion.

As already mentioned, any clock may be employed to define the unit of time.

E.g., one can resort to the daily motion of the planet, i.e., to eqn. (B.21").

The second (unit of time) may be defined as the 86 400-th part of the average solar day. One refers to average solar day because there are some deviations from linearity in eqn. (B.19) that relate $\tau_{i\alpha}$ and $\tau_{j\alpha}$; this equation takes an idealized form which is the more exact, the higher $\Delta\tau_{i\alpha}$ (and implicitly $\Delta\tau_{j\alpha}$). It should be recalled that the linearity of functions that relate the variations of equivalent timing parameters is often approximate (because the equivalence of clocks is approximate).

Substitution of $\Delta\tau_{j\alpha}$ with 2π radians in eqn. (B.21"), according to the definition, yields Δt_{α} = 86 400 seconds, i.e.,

$$2\pi = v_{j\alpha} \cdot 86\ 400;$$

$$v_{j\alpha} = \frac{2\pi}{86\ 400} = \frac{\pi}{43\ 200} \text{ rad/s.} \qquad \text{(B.24)}$$

Thus one second is the time interval required by Earth to rotate about its axis through an angle $\pi/43\ 200$ radians.

The way the multiples and submultiples of the second are defined is evident.

E.g., an observer notices that a certain space s_0 was travelled over by a point-like moving body M in 10 minutes, which is equivalent to establishing that s_0 travelled by M corresponds to a rotation of $\phi_0 = 10 \times 60 \frac{\pi}{43\ 200} = \frac{\pi}{72}$ rad of the Earth about its axis.

A more modern definition (equivalent to the one just given) of the second will be presented below; it is related to the irreversibility of time and the cyclic (periodic) character of some motions.

Once the manner for the complete expression of time (variable originating in the synthesis of variations of equivalent timing quantities) by means of a unit is available, any motion, that is, any variation of a state quantity may be expressed as a function of time.

According to eqn. (B.17), a timing parameter $\tau \in L_\alpha$ (the subscript α attached to τ is to emphasize that τ belongs to the L_α equivalence class and that in this class there are many, ν_α, such clocks) is related to time by the equation $\tau = vt_\alpha + k$, where t_α is the time of the equivalence class L_α. This aspect of time associated with an equivalence class of the set of clocks, is now once more emphasized.

Consequently, the functions (B.13') become:

$$q_{is} = \phi_{is}(\tau) = \phi_{is}(vt_\alpha + k) = g_{is}(t_\alpha), \tag{B.25}$$

where $i = 1, 2, \ldots, \gamma_s$ and $s = 1, 2, \ldots, n$.

The ensemble of functions $q_{is} = g_{is}(t_\alpha)$ with $i = 1, 2, \ldots, \gamma_s$ form the law of motion of the system Γ_s (whose state quantities are q_{is}) expressed with respect to the time t_α.

For the observer O, who has employed clocks in the equivalence class L_α to measure and express the motion of systems Γ_s whose transformation is studied, all states corresponding to the moment $t_\alpha = t_{p\alpha}$ are simultaneous, characterized by values $q_{pis} = g_{is}(t_{p\alpha})$ of state quantities, and form the present at time $t_\alpha = t_{p\alpha}$.

All states corresponding to a value $t_\alpha < t_{p\alpha}$ represent the past with respect to the moment $t_\alpha = t_{p\alpha}$, and those for which $t_\alpha > t_{p\alpha}$ represent the future with respect to $t_\alpha = t_{p\alpha}$. This results from the increasing of the timing parameter with time. Once more we specify that no cyclic variation (any variation is non-cyclic or cyclic and periodical only when referred to another variation) of certain timing parameters has been considered here yet. Only those parameters have been considered whose mutual variation is monotonically increasing, like their linearly increasing time-variation according to eqn. (B.17).

Note. Time was defined as a variable, a quantity which forms a continuous set of values, with abstract character resulting from the synthesis of all timing parameters of an equivalence class. Because a synthesis of timing motions belonging to one equivalence class is achieved and a certain equivalence class is chosen for synthesis, on generalizing and abstracting the following questions arise: (a) May any class of equivalent clocks be chosen? (b) What eventual criteria of selection of one class may be adopted?

However, it is important to emphasize that the basic notion of time, this entity with such deep implications in knowledge and thinking, may not be regarded as a continuous set of values, even following a synthesis of certain quantities (timing parameters) designed to achieve ordering of states, simultaneity and succession, i.e, the general image of motion for any observer.

It is true that quantitative expressions in the calculations required by the science of nature employ this variable as a mathematical variable whose characteristics stem from the way it was introduced.

It is important to understand that time is much more than just the variable t_α which represents it.

Time is the measure of general motion and the variable that denotes it is only a manner of expressing this general motion whose content has to be studied in all its considerable depth.

For an observer that observes no motion (transformation) the concept of time loses sense (time "disappears").

B.I.3. The properties of time

The essence of this entity called time, the way it was conceived, its characteristics, the consequences for the study of nature, and its implications in general practice represent points of vital importance that any basic approach to the laws of being, of their understanding and utilization, have to consider.

Obviously, when one reaches subtle interpretations there is not just the possibility but even a high likelihood that different points of view arises. Thus, there are opinions that oppose the tendency to <u>spatialize</u> time, to treating it as a fourth dimension of space. It is considered that such an interpretation may be accepted with an <u>exclusively formal</u> character in calculations, for elegant and homogeneous mathematical expressions. But the physical essence of time may not be identified with that of space.

Obviously this point of view is not expressed for the first time, but this additional emphasis associated with the arguments already presented and the ones that follow is welcome in a book which attempts to treat the problem with fair accuracy and correlate it with the theory of relativity, i.e., with that theory which has encouraged the tendencies to spatialize time. What we undertake is an interpretation of the concept of time which rejects the image of time as a fourth dimension of space (an image accepted only in the mathematical formalism and not in a physical sense); however the interpretation must not contradict the results of the theory of relativity, because the intrinsic (relativistic) space-time relation must not be pushed so far as to mean an intimate identification of space and time.

B.I.3.1. The irreversibility of time. A property of time of major importance, which distinguishes it from space and awards it a physical sense distinct from any space dimension, is its irreversibility.

It is proper to mention that this is not the only basic distinction between time and space. Time is a measure and consequence of motion. It is true that space, as time, may not be conceived outside matter (substance, energy, field) and matter in its turn is necessarily associated with motion (in the general sense of transformation). The interdependence between the basic entities and motion is rigorously expressed by relativistic physics.

However, leaving aside the irreversibility of time, there is another essential difference between time and space. Any variation of a space coordinate during a motion implies variation of a time coordinate without assuming necessarily the variation of another space coordinate (coordinates).

Likewise, a space (a length) may be measured by comparison with another standard space (another length), but no relative motion of them is required. Without the perception of a motion time may not be measured.

Let us return now to irreversibility.

Any moment has to characterize for the observer a state of the Universe which assumes an ensemble of values for all state quantities. If all states are different (differing from one another by at least a state quantity of at least one system), all moments are expressed by distinct values. However, the set of values which represent the moment is precisely the set of values of a time variable. The fact that all states are different requires a variation of time in which its values do not repeat. This is a monotonic variation admitted to be monotonically increasing, which denotes the irreversibility of time.

The reversibility of time would have to assume the existence of at least two identical states of the universe noted by the same observer which have between them at least a state different from the two identical ones. This is not possible, not as long as in nature there is at least one irreversible transformation (as, e.g., the variation of entropy); even if one assumes that none of the state quantities noted by the observer represents an irreversible phenomenon, the probability that some general states of the whole universe reproduce will decrease, as the number of state quantities observed (contained in the universe of the observer) increases, and in the end become zero when the number of such quantities is infinite.

Therefore, one can hypothetically admit that time is reversible only for an observer existing in an isolated microworld (who notes nothing except this microworld) in which only a small number of state quantities vary, of which no one represents an irreversible motion in a unique sense.

However such a universe and an observer are only

strictly hypothetical.

Moreover, such an observer should ignore previous experience, that is, previous observation in a more complex universe in which time, for reasons explained above, appears as irreversible.

Thus, if the observer holds previous information from a complex universe with irreversible time (at least probabilistically), this information should not be stored, so that the microworld where time may seem reversible is sensed by the observer as such. All these conditions are plain imagination. Thus time will be considered as it is in fact, i.e., irreversible.

In such a microworld, if the variations of all (relatively few) state quantities are quantized, the variation of time originating in a discontinuous succession of states seems discontinuous.

However a time that varies discontinuously could be recorded only under conditions similar to those that accredit the idea of irreversible time which are unacceptable; these are: a microworld (with a finite number of state quantities all with quantized variation) and an observer who had not created a world model based on observations (informations) previously recorded.

In fact, time varies continuously, as pointed out and as it has always been considered.

Now, at this point, it is appropriate to analyze the eventual existence of clocks with periodic motion or clocks with discontinuous motion (which, e.g., give discontinuous signals or pulses).

Consider therefore a periodic motion and let q denote a state quantity of system Γ whose motion was considered constant when measured and described by an observer O who resorted to the equivalence class L_α of clocks and who obtained the time variable t_α for the ordering of states and the establishment of their succession, i.e., to form an image of the motion.

With the timing parameter τ of clock $C \in L_\alpha$ (it is known that $\tau = vt_\alpha + k$) and resorting to eqn. (B.25), one can write:

$$q = \phi(\tau) = \phi(vt_\alpha + k) = g(t_\alpha). \qquad \text{(B.25')}$$

Stating that q varies periodically amounts to saying that there is a time interval Δt_α whose value (measured) is constant such that q as well as its n-th derivatives (n = 1, 2, ...) satisfy the conditions

$$\begin{aligned} q(t_\alpha) &= q(t_\alpha + \Delta t_\alpha), \\ q^{(n)}(t_\alpha) &= q^{(n)}(t_\alpha + \Delta t_\alpha), \end{aligned} \qquad \text{(B.26)}$$

whatever the moment t_α considered.

Now, if function $q(t_\alpha)$ is n times differentiable, the derivatives $q^{(n)}$ are to be taken in a classical sense. In the opposite case (when $q \notin C^n$), the derivatives have the generalized sense of the theory of distributions.

This interval $\Delta t_\alpha = T_\alpha$ is called the <u>period</u>.

The observer O notes a constant value of the interval, called the period, by means of the timing parameter in terms of the known relation:

$$\Delta\tau = v\Delta t_\alpha = vT_\alpha,$$

which is associated with (B.25') and (B.26).

Indeed:

$$\phi\ (\tau) = g\ (t_\alpha) = \phi\ (vt_\alpha + k) =$$

$$g\ (t_\alpha + T_\alpha) = \phi\ (vt_\alpha + vT_\alpha + k) =$$

$$\phi\ (\tau + \Delta\tau) \tag{B.27}$$

and

$$\frac{d^n}{dt_\alpha^n} g\ (t_\alpha) = \frac{d^n}{dt_\alpha^n} \phi\ (vt_\alpha + k) =$$

$$v^n \frac{d^n\phi}{d\tau^n};\ n = 1, 2, \ldots . \tag{B.28}$$

Thus:

$$v^n\ \phi^{(n)}\ (\tau) = v^n\ \phi^{(n)}\ (\tau + \Delta\tau). \tag{B.29}$$

That is, in fact, the observer measures the period $\Delta\tau$ given by the timing parameter to be then translated into the period T_α.

This is done by means of the constant v which may, in particular, have the value unity (but not without dimension).

Taking into account the irreversible monotonic variation (continuously increasing) of time, the timing parameter $\tau = vt_\alpha + k$ may never vary periodically.

Once the time is defined by some non-periodic motions employing the monotonic variation of some state quantities (which satisfy the conditions required to be timing parameters), periodic variations may be employed (the periodicity assumes the reversibility of variation) to measure and express time but only correlated (explicitly or implicitly) with other quantities which vary linearly with time and thus transform irreversibly at any given moment (implicitly non-periodic).

This is achieved by measuring time at any origin (any event) by the number of cycles (periods) completed from that time origin and the value of the periodic variable at the given moment.

Obviously, the period of cyclic variation is assumed constant.

The procedure is similar when one deals with signals (pulses) given at equal time intervals.

It is worth noting that, although these clocks with cyclic variation (continuous or discontinuous) may be employed to measure time, they may not underlie the definition of time as a general measure of motion unless associated with other monotonic variations or, which is the same, associated with the number of cycles completed (from the time origin).

Sometimes the association of periodic (cyclic) variations with a monotonic non-periodic one is intrinsic. In this case the system has a periodic motion only in respect of some of its variables. This is the case, e.g., for the propagation of a wave. The oscillation induced by the wave is associated intrinsically with the propagation of the wave front, a motion that may be non-periodical.

Even the second may be defined as the duration of 9 192 631 770 periods of radiation corresponding to the transition between the two hyperfine levels of the ground state of the caesium-113 atom.

To measure one period one has to resort initially to a non-periodic timing motion.

Finally, it is noteworthy that in a <u>fictitious</u> four-dimensional space in which the fourth dimension, time, is associated with the other three of physical space, motion of any moving body may never be represented by a closed curve due to the irreversibility of time.

Indeed, consider a point M and an infinite and continuous set of events E (x, y, z, t) which consist of the transit of the point M in its motion, referred to a frame Oxyz, through positions of coordinates x, y, z at times t.

In Minkowski space this set η is represented by a curve C_η whose points are just the events E (x, y, z, t).

Whatever the motion of point M, if <u>any</u> pair of events E_A (x_A, y_A, z_A, t_A) and E_B (x_B, y_B, z_B, t_B) is considered, these events being positions of M of space coordinates x_A, y_A, z_A and x_B, y_B, z_B at times t_A and t_B, respectively, reached during motion, then it is possible that $x_A = x_B$ or $x_A \neq x_B$, $y_A = y_B$ or $y_A \neq y_B$, $z_A = z_B$ or $z_A \neq z_B$. But it is compulsory, due to the irreversibility of the variation of time, that $t_A < t_B$ (if E_A is anterior to E_B) or $t_A > t_B$ (if E_A succeeds E_B); anyway $t_A \neq t_B$ and therefore $E_A \not\equiv E_B$.

Obviously, it is understood that in the succession of events E_A and E_B there is a set of events E which succeed one and precede the other, i.e., the particle M travels from E_A to E_B or in opposite sense.

This amounts to saying that there is an arc of a curve C_η of non-zero (measured) length which has one end in A and another one in B. Likewise, if the trace of M in Oxyz is the curve C_0 on which points M_A (x_A, y_A, z_A) and M_B (x_B, y_B, z_B) lie, the arc $M_A M_B \in C_0$ is assumed to be of a length different from zero, even though, in particular, $x_A = x_B$, $y_A = y_B$, $z_A = z_B$, i.e., $M_A = M_B$, which means that the arc $M_A M_B \in C_0$ is a closed curve in the three-dimensional physical space Oxyz.

However, because $E_A \neq E_B$ at <u>any</u> instance, it follows that in the Minkowski space the curve C_η is necessarily open, because $t_A \neq t_B$.

It is considered useful that, when the presentation of principles ends, examples are supplied related to the variation of time, its irreversibility, to the periodicity of some motions associated with this irreversibility, to the continuity of moment sets, and to the curves representing the motion of one single material particle in Minkowski space, which have to be open.

It is worthwhile keeping in mind that, essentially, time expresses a major truth: in its general motion the universe does not have identical states and, moreover, does not assume a succession of states identical with others belonging to the past.

Repetition would mean returning in time (i.e., to an identical state for the <u>whole</u> universe, in its micro-, macro- and mega-cosmic aspects regarding all material systems).

And if the states of the universe (identical with those that preceded the present) were reproduced in inverse order (likewise, in the whole universe), this would mean a variation of time <u>towards the past</u>.

As has been seen, all these are absolutely impossible because, firstly, there are irreversible phenomena (which at least rule out the reproduction of states of some material systems), successions of transformations which are deterministically conditioned and, secondly, even without taking these into account, the infinite number of state variables of the infinity of existing material systems forbid, in probabilistic terms, the repetition of a state (more so of several states) of the whole universe.

This is the way one should understand the irreversibility of time, as a measure and a consequence of general motion, of a continuous succession of states of the universe.

B.I.3.2. The velocities and accelerations in various equivalence classes. In the foregoing the idea was accredited that the concept of time in the synthesis, generalization, and abstraction of the variation of timing parameters belonging to the same equivalence class, i.e., of those timing parameters which, being mutually equivalent, undergo variations related by linear functions according to (B.16) and allow, as a consequence, the introduction of time by eqn. (B.17). That is, with the known notations, one can write the equations:

$$\tau_{i\alpha} = a_{ij\alpha}\,\tau_{j\alpha} + b_{ij\alpha};$$

$$i = 1, 2, \ldots, \nu_\alpha;\; j = 1, 2, \ldots, \nu_\alpha;$$

$$\tau_{u\alpha} = v_{u\alpha}\,t_\alpha + k_{u\alpha};$$

$$u = 1, 2, \ldots, \nu_\alpha,$$

where t_α is time given by the ν_α clocks $C_\alpha \in L_\alpha$ (L_α being an equivalence class, ν_α may be finite or infinite).

We know that between the variations of (timing) quantities $\tau_{i\alpha}$ and $\tau_{j\alpha}$ may or may not exist causal relationships. The observer O is aware of the (bi-univocal) correspondence between sets $\theta_{i\alpha}$ ($\tau_{i\alpha} \in \theta_{i\alpha}$) and $\theta_{j\alpha}$ ($\tau_{j\alpha} \in \theta_{j\alpha}$), regardless of the possible existence of deterministic relationships between them, which are irrelevant for this stage of our study.

The rate of variation ω_{is} of a state quantity q_{is} which characterizes the state (and whose variation contributes to characterization of the motion) of the system Γ_s with respect to the variation of the timing parameter τ is, according to the dependence expressed by eqn. (B.13'), given by the equation:

$$\omega_{is} = \frac{dq_{is}}{d\tau} = \frac{d}{d\tau}\,\phi_{is}(\tau), \tag{B.30}$$

and the acceleration Ω_{is} of q_{is} with respect to τ is:

$$\Omega_{is} = \frac{d\omega_{is}}{d\tau} = \frac{d^2 q_{is}}{d\tau^2} = \frac{d^2}{d\tau^2}\,\phi_{is}(\tau). \tag{B.31}$$

In particular, q_{is} may also be a timing parameter. And even more particularly, the velocity of variation of a timing parameter $\tau_{i\alpha}$ against another equivalent $\tau_{j\alpha}$ (with $\tau_{i\alpha} \in L_\alpha$, $\tau_{j\alpha} \in L_\alpha$) is:

$$\omega_{ij} = \frac{d\tau_{i\alpha}}{d\tau_{j\alpha}} = a_{ij}, \tag{B.32}$$

and the acceleration satisfies:

$$\Omega_{ij} = \frac{d^2\tau_{i\alpha}}{d\tau_{j\alpha}^2} = 0. \tag{B.33}$$

The velocities and accelerations with respect to time are (for any state variable q_{is}):

$$v_{is\alpha} = \frac{dq_{is}}{dt_\alpha} = \frac{dq_{is}}{d\tau_{u\alpha}} \frac{d\tau_{u\alpha}}{dt_\alpha} = \omega_{is\alpha} v_{u\alpha},$$

$$a_{is} = \frac{d^2q_{is}}{dt_\alpha^2} = \frac{dv_{is\alpha}}{dt_\alpha} = \frac{dv_{is\alpha}}{d\tau_{u\alpha}} \frac{d\tau_{u\alpha}}{dt_\alpha} = \tag{B.34}$$

$$\Omega_{isu\alpha} v_{u\alpha}^2,$$

with obvious notations (i.e., those employed above with indexes such that they emphasize the state quantity, the parameter with respect to which the velocities and accelerations are considered, and the equivalence class L_α employed to introduce the time variable).

Obviously, $\nu_{u\alpha}$ is constant.

The dimensions of velocities and acclerations with respect to parameters τ or time t result immediately from eqns. (B.30), (B.31), (B.32), (B.33), and (B.34) and the dimensions (units) of quantities involved (q_{is}, τ_{is}, τ_{js}, $\tau_{u\alpha}$, t_α).

The derivatives of these equations have the classical meaning if the functions are differentiable.

In case their differentiability does not hold throughout the domain of interest, the relations above continue to stand, but the operators of derivation are considered in the sense of the theory of distributions.

Note. Because the functions in (B.13'), in the construction of the equivalence classes in which the set of clocks is divided (implicitly, the relation between the timing parameters) may depend on the findings of the observer, i.e., may differ from one observer to the other, the velocities and accelerations considered above may have a relative character, depending on observations (i.e., on measurements carried out). Admitting the ideal case (the most profitable possible) of the absence of any measurement error, the observations (measurements) depend on the relative state of motion between the observer and the phenomenon studied. This, in principle, is due to the fact that information, i.e., the image of motion (transformation) reaches any observer by means of an agent which carries information (this aspect is studied next). This is the physical essence, the cause of the relativistic (Einsteinian) character of velocity and acceleration of motion in general.

B.I.3.3. The transport of information. All throughout this book, especially in its first chapter regarding the theory of relativity, a lot was said about the way, well-known in modern physics, that the influence of one body upon other bodies, i.e., the mutual influence (interaction) between material systems, propagates and that at the same time information propagates from one system to the other.

The fact that two (or more) systems may not communicate, i.e., exchange information, except through the existence of some form of energy which travels from one system to the other, i.e., due to some transport of information, has become an elementary truth. The presence of this truth in considerations related to the discovery of nature is one of the main aspects which differentiate modern science from classical (pre-relativistic) science.

Any observer O observes the motion of a material system Γ, the variation of its state variables q_i ($i = 1, 2, \ldots, \gamma$), by means of other motions which carry information from Γ to O.

Supposing that there are various agents which carry information from Γ to O (e.g., electromagnetic waves, acoustical waves, electric current, and thermal flux), the observer O traces the motion of Γ by means of that agent which, chronologically speaking, brings the first item of information.

Thus, the observer O perceives, by means of electromagnetic waves, an event E occurring at point M and time t_1. Obviously it is assumed that timing is done with a clock. If there also exists an acoustical effect of E, let t_2 be the moment O notes this effect. Obviously $t_2 > t_1$. If E has also a thermal effect assume this gets to O at time $t_3 > t_2$.

The observer O notes the event E as occurring at time t_1 when the first item of information is received.

Therefore any observer acquires a general and detailed image of the universe; the observer estimates motions (transformations), establishes their laws, both qualitative and quantitative, finds the correspondence of state variables, that is, notes the states of the universe and with them their succession; consequently, the observer introduces the notion of <u>time</u> depending on the first information received, that is, on the information carried by the <u>fastest information</u>

carrying agent. This is light or, in general, the electromagnetic wave.

From a physical point of view, the phenomenon occurs as follows:

A certain motion, in a point of space, causes a set of changes of emissions (or reflections) of electromagnetic energy (or energy of another nature). This propagates in space to the observer. That is the only way information gets to the observer.

Thus, a state of motion may be noted only if there is another motion, i.e., that of the information carrier, as the motion of an energy field (the electromagnetic field), from whose motion an observer (in the field) infers the motion (or absence of motion) of the material system in question (which has emitted or reflected the electromagnetic wave).

In fact, any observer comes in direct touch only with the motion and changes of this agent that carries information. The importance of this fact is obvious.

The classical, pre-relativistic science has not considered this "go-between" which carries information, neither did it consider the effects its existence exerts upon information, that is upon the image that the observer forms, upon his knowledge of the universe, in particular upon the succession of states and implicitly the concept of time.

Therefore, the universal time, or the Newtonian time which is not influenced by anything, independent of any transformation, with its perpetual and uniform elapse is the fictitious, imaginary time resulting from an idealized scheme, from a model of nature which lacks an aspect of considerable importance: the transport of information, the motion in space of perturbations which impinge on the observer and create the image of their causes - the transformations (motions) observed. To remove this information transport phenomenon from the general model of the universe is equivalent to considering that information travels at an infinite speed, which is specific to pre-relativistic science.

This model has successfully faced the confrontation with experiment so long as the techniques and finesse of experimenting were of such a level that they could not give evidence of small errors brought about by disregarding the phenomenon of the transport of information.

However, experiments carried out with advanced techniques and higher accuracy will always furnish evidence of these errors. The higher the speed of information in comparison with the velocity of the transformations (motions) that the information refers to and the motion of the observer, the smaller these errors will be.

The way the information transport by electromagnetic waves interferes and the role played by the speed of light in the derivation of relations of relativistic kinematics and dynamics are most convincing and were presented in the first part of this book - which dealt with relativistic mechanics.

Now only those aspects are investigated that are insufficiently treated or ignored in the specialized literature.

Regarding the synchronization of clocks (the notion of synchronized clocks is known from the theory of relativity and discussed in the first part of this work), we recall here for easy understanding that the clocks of a frame $Oxyz$ are so synchronized that all events whose information is carried by the same wave front appear as simultaneous for any observer who receives the information from that wave, no matter what the position of the observer or the location in space where the events in question occurred, provided that all observers are at rest with respect to the same frame (in which synchronization of clocks was achieved).

That is, if at various points of space M_1, M_2, ..., M_n there occur the events E_1, E_2, ..., E_n and if, from the relative positions of points M_i ($i = 1, 2, ..., n$) and from the time coordinates of events E_i ($i = 1, 2, ..., n$), one achieves the coincidence of any event E_i and the transit through M_i of the front of one electromagnetic wave, then the wave front carries along the information of all events E_i it passes over to all observers it touches, and as a consequence for all these observers the events E_i appear as <u>simultaneous</u>.

Consider now p such observers O_k ($k = 1, 2, ..., p$) and a series of events η_{jk} ($j = 1, 2, ...$).

It is specified that observers O_k are at rest with respect to one frame Oxyz and that any of the events η_{jk} consists of the reception by observer O_k of a set of l_j events E_{ji} ($i = 1, 2, ..., l_j$) whose information is carried by the same wave front, i.e., which are simultaneous for observers O_k.

Let t_{jk} be the moment the event η_{jk} occurs for observer O_k.

And define the time intervals:

$$\Delta t_{j+1,j,k} = t_{j+1,k} - t_{jk}, \qquad \text{(B.35)}$$

delimited for observer O_k by events η_{jk} and $\eta_{j+1,k}$.

Because, for all observers O_k, the events η_{jk} are carried by the same electromagnetic wave front, if the frame Oxyz is inertial and the clocks used by observers O_k are synchronized, the time intervals $\Delta t_{j+1;j,k}$ have the same measure for any observer O_k, that is, are independent of the subscript k:

$$\Delta t_{j+1;k} = \Delta t_{j+1,j}; \qquad k = 1, 2, \ldots, p. \tag{B.36}$$

On the other hand, the synchronization of clocks is possible only if they belong to the same equivalence class.

The statement is obvious if one takes into account the way it was assumed that the synchronization of clocks is achieved in the theory of relativity and the definition of the equivalence classes of the set of clocks. And, in order that eqn. (B.36) exists whatever the relative positions of observers O_k and the direction of wave fronts which carry the informations regarding the events η_{jk}, it is necessary that the propagation of a light wave is a motion which, as a timing phenomenon, belongs to the same equivalence class with synchronized clocks of the observers O_k.

All these could have been formulated briefly and simply as follows: synchronization of clocks in an inertial frame assumes that all clocks belong to the equivalence class to which the propagation of electromagnetic wave (taken as a timing phenomenon) belongs.

The affiliation with the same equivalence class of all clocks employed by observers O_k yields:

$$\tau_k = v_k t + b_k;$$
$$k = 1, 2, \ldots, p, \qquad \text{(B.37)}$$

where τ_k is the timing parameter of the clock observer O_k uses, t the time resulting from the synthesis and abstraction of clocks that belong to the equivalence class, including the clocks of observers O_k (that is the time of the equivalence class in question), and v_k and b_k are constants with obvious significances.

Inclusion in the same equivalence class L of clocks C_k of parameters τ_k and of the motion of propagation of the electromagnetic wave (compelling in order that synchronization of clocks is possible) yields immediately:

$$\tau = ct + b, \qquad \text{(B.38)}$$

where τ is the space travelled by the wave front, c the speed of the wave (light), and b a constant (t is obviously the time of the equivalence class in question).

A conclusion arises:

In the reference frame where synchronized clocks are equivalent with the motion (propagation) of light, this speed of propagation is constant. Such reference frames are referred to as inertial.

Note. All throughout these considerations, light (electromagnetic wave) was considered the information carrier. This is so because, from the findings so far, it follows that light is the fastest information carrier and therefore gives the observer the image of phenomena.

In principle, whatever was and is going to be said regarding light as an information carrier may be extended to any other information carrier.

One can thus imagine a theory of relativity in which the speed of light is replaced by the velocity v_0 of any other information carrier.

However taking into account the relativistic equations, such a change of information carrier precludes expression of any motion of velocity $v > v_0$, because the quantities $\gamma = \sqrt{1 - \frac{v^2}{c^2}}$ would become $\gamma_0 = \sqrt{1 - \frac{v^2}{v_0^2}}$ which loses real sense for $v > v_0$. Thus replacement of light with another agent that carries information is possible only if its velocity, v_0, is not exceeded by another velocity, i.e., $v_0 \geqslant c$. But such an agent whose velocity is higher than that of light has not been found so far, either in Nature or in technology.

Therefore, light (an electromagnetic wave) continues to be considered the information carrier, as relativistic physics does.

B.I.3.4. Information transport as a timing motion. First of all a simple observation or, rather, a resumption of a statement which has already been made:

All the clocks which belong to the same equivalence class as the propagation of light (also considered a timing motion) may be synchronized and the frame the timing motions are referred to, or where the timing transformations occur, is an inertial frame.

The time variable has to be under these conditions proportional to the space travelled through by the front of the light wave, and the proportionality factor is the speed c of light. Such a time is the light time.

This reasoning may immediately be extrapolated.

Consider an information carrier A, in particular a light wave.

The motion in space of this information carrier, taken as a timing phenomenon, decides the equivalence class, which, by abstraction and general synthesis of all clocks equivalent to A, leads to the notion of time.

This is the time of the information carrier denoted by A, in particular of light itself.

Any interval Δt has to be proportional to $\Delta\tau$, the space travelled by the information carrier.

In fact, Δt is a variation of a fictitious variable. A physical reality is $\Delta\tau$ and so are all the variations of physical quantities that the observer notes.

Let them be $\Delta\tau_k$ if the timing parameters are involved or Δq_{is} if some state variables are involved.

To allow synchronization of clocks, they have to belong to the same class and, as already seen:

$$\frac{\Delta\tau_k}{\Delta\tau} = v_k \,, \tag{B.39}$$

and v_{kl} are constants measured by the observer.

Moreover, with the space τ covered by information, the functions:

$$\tau_k = f_k\,(\tau). \tag{B.40}$$

are such that:

$$\frac{d\tau_k}{d\tau} = \frac{d}{d\tau}\,f_k\,(\tau) = v_k \,, \tag{B.41}$$

$$\frac{d^2}{d\tau^2} f_k (\tau) = 0,$$

which results convincingly from the above.

The fictitious variable t is in fact chosen arbitrarily so that eqn. (B.38) is obeyed.

The definition of the unit which allows a quantitative expression of time t may be given, starting directly from the propagation of information or, as it was usually done so far, starting from another timing phenomenon belonging to the same equivalence class with the propagation of information, even though this fact was not thought and expressed so far.

Because synchronization of clocks (in the sense given by the theory of relativity) is possible only if the clocks are equivalent with the clock based on the motion of the information carrier (and thus mutually equivalent), the introduction of the notion of time amounts to comparing any transformation (motion) with that of the information carrier. Although it is not compelling at all, it is usual to introduce that abstract variable called time which varies linearly with the timing parameters in that equivalence class, that is, at first, with the space travelled through by information.

A return to the primary phenomenon should allow the expression of all state quantities by functions of the form:

$$q_{is} = \phi_{is} (\tau), \tag{B.42}$$

where τ is the space travelled by the information carrier (in fact by light).

The velocities and accelerations should be expressed as functions of this parameter τ according to (B.30) and (B.31) but, according to the justification presented above, only motions whose velocity is lower than that of information may be studied; therefore the information is carried by the highest speed motion and it follows that $\omega_{is} < 1$ (B.30), that is, all velocities divided by the speed of information yield subunit values.

Return now to the abstract variable called time, which represents (as mentioned) a concept which is much more complex and subtle than that of simple variation utilized in calculation, i.e., that of the general measure of the infinite complex of motions.

Obviously this variable called time has to be correlated intrinsically with the propagation in space of the general influence of information. (Freezing the propagation of information, one is no longer able to note the motion, time disappears.) Formally time appears as a simple variable proportional to the space travelled by information (influence). One should analyze the factor of proportionality which qualitatively relates the variation of space travelled by information to the variation of the abstract variable called time, i.e., in fact the constant c in eqn. (B.38), $\tau = ct + b$, with the known significance of symbols.

Obviously the value of c results from the selection, which is arbitrary, of the variable t, more precisely of its unit because t has to satisfy the unique condition expressed by eqn. (B.38). In case time is expressed in seconds, c represents the space travelled along by the light wave (or more generally, by information) during 9 192 631 660 periods of the radiation corresponding to the transition between the two hyperfine levels of the ground state of caesium-113 atom (according to the definition of the second).

Evidently, one is aware of the errors of measurement, however small. They do not matter as one is interested only in the fundamental aspect of the problem.

Now, the core of an aspect of outstanding importance is reached: the space covered by light (or, in general, by information).

This assumes the existence of a frame, a reference frame. A basic question arises: how does the velocity of light or information, c, behave when one changes the reference frame?

This question is not answered axiomatically (as Einstein did) but by an analysis which justifies the truth.

The reasoning is based on the following observation: when the relationship between distance $\Delta\tau$ travelled through by the wave front (light, information) and another timing motion, designated to define the time unit (e.g., the second), is considered, the observations, experiments, or measurements which establish this relationship also assume an information transport achieved by light (or, more generally, by the agent which does the job). It has been very often reiterated that any communication, any experiment or measurement is performed only with the help of an information carrier. Therefore, its role (the role of electromagnetic wave) becomes a major one, especially associated with the fact that only clocks belonging to the same equivalence class (which contains the motion of propagation of information that in itself is considered a timing motion) may be synchronized (in the sense of the theory of relativity).

The following is a resumption, in a slightly different form, of some reasoning characteristic for the theory of relativity presented in the first chapter.

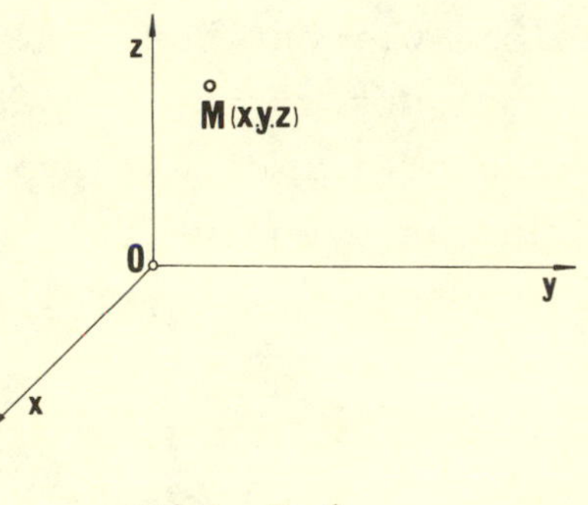

Fig. B-1

It is assumed that at any point M (x, y, z) of the frame Oxyz a clock is placed (Fig. B-1). (Obviously, this is only theoretically possible.) Assume all the clocks are synchronized whatever the position of M, and let t_{1M} be the moment the observer at M notes an event E_1 employing the clock at M and t_{2M} the moment the same observer notes event E_2 (employing the same clock).

At the origin of the reference frame, i.e., at point O, there is another observer and a clock and events E_1 and E_2 correspond to moments t_{01} and t_{02}.

Synchronization of clocks ensures that, whatever the position of M:

$$t_{02} - t_{01} = t_{2M} - t_{1M}. \tag{B.43}$$

This amounts, in fact, to stating that all synchronized clocks in Oxyz show that all observers in

Oxyz (whatever the position of M) may use only one clock, e.g., the one at O, and operate a simple shift: that corresponding to the time required by information to cover distance OM (the shift of clock origins which would entail $t_{02} = t_{2M}$ and $t_{01} = t_{1M}$).

Essentially, this is the synchronization of clocks. It would be impossible without eqn. (B.43), which assumes the timing motions of the transport of information and the motions of all clocks synchronized in Oxyz belonging to the same equivalence class, because disobeying of some proportionality relationships would cause variation of the $t_{2M} - t_{1M}$ intervals with the position of M.

Consider now two frames in relative motion, Oxyz and Aαβγ. All points M in the space at rest with respect to Oxyz form the space Oxyz and all points P at rest with respect to Aαβγ form the space of the frame Aαβγ. Obviously points M and P are in relative motion.

We assume the motion of the two frames is such that the clocks in both of them belong to the same equivalence class with the motion of light waves (the agent that carries information), which is a clock by itself, whatever the position of clocks. If, under these circumstances, clocks in each frame may be synchronized, i.e., an observer attached to Oxyz in M uses the clock at O and notes an interval Δt_0 between two events, the same as the one found in the frame attached to M (to the observer), and an observer at P attached to Aαβγ notes a time interval Δt_A between two events, either using the clock at A or one at P (attached to the observer), then the two reference frames are in inertial motion.

Indeed, such a situation assumes synchronizing of clocks based upon their belonging to the same equivalence class with the motion of information (light), while the systems where this is possible are referred to as inertial. In such frames eqn. (B.43) is obeyed.

Thus, it was admitted that Oxyz and $A\alpha\beta\gamma$ are inertial frames. The relative motion of such systems is inertial.

In previous considerations it was suggested that by a clock in a certain frame (e.g., Oxyz) is understood a motion referred to that frame. And by a clock located at a certain point, e.g., M, is understood either a (non-mechanical) transformation occurring at the point M, assumed fixed with respect to the frame in question (as, e.g., Oxyz), or a mechanical motion in the neighbourhood of the point M.

It was assumed that Oxyz and $A\alpha\beta\gamma$ are in relative motion. Oxyz is considered the basic frame (inadequately called fixed because one can do the opposite assuming $A\alpha\beta\gamma$ fixed and Oxyz mobile), the motion between the two frames being in this case the motion of $A\alpha\beta\gamma$ with respect to Oxyz.

Consider P (α, β, γ) belonging to $A\alpha\beta\gamma$, that is, at rest with respect to $A\alpha\beta\gamma$.

Consider also two events E_1 and E_2.

For the sake of simplicity, consider the case when the two events occur at P at times t_{P1} and t_{P2}, as noted by the observer at P on the clock at P, and times t_{A1} and t_{A2} on the clock at A. According to (B.43), because of synchronization of clocks, one can write:

$$t_{A2} - t_{A1} = t_{P2} - t_{P1} = \Delta t_P. \qquad \text{(B.43')}$$

Due to the motion of $A\alpha\beta\gamma$ with respect to Oxyz, the point P moves with respect to frame Oxyz. It coincides at any moment with a fixed point M in Oxyz, and the continuous set of these points M forms the trace (trajectory) of P in Oxyz: a curve D (in fact a linear trajectory because P is at rest with respect to $A\alpha\beta\gamma$ and this latter is inertial, as is Oxyz). Concomitantly with events E_1 and E_2, P occupies the positions M_1 (x_1, y_1, z_1) and M_2 (x_2, y_2, z_2), respectively.

Obviously $M_1 \in D$ and $M_2 \in D$.

Assume now that the observer attached to O does not employ the observer's clock or that at A (with which that at P is synchronized) but clocks at points $M \in D$ reached by P in its motion against Oxyz. Obviously these latter are synchronized with each other and with the clock at O. Given the relative motion of frames $A\alpha\beta\gamma$ and Oxyz, synchronization of clocks of $A\alpha\beta\gamma$ and Oxyz does not allow one to assume the eventual synchronization or non-synchronization of clocks in $A\alpha\beta\gamma$ against those in Oxyz, and conversely.

On the clock of M_1 the observer in P notes the value t'_{M1} when E_1 occurs (and P lies at position M_1 in Oxyz), and, when E_2 occurs, the clock at M_2 shows to the observer at P (who now coincides with M_2) the time t'_{M2}.

Thus, when the observer's own clock (at P) is employed, the time interval between the <u>same</u> events E_1 and E_2 is Δt_A (given by (B.43')), while the clocks in Oxyz give a time interval:

$$\Delta t'_0 = t'_{2M} - t'_{1M} = t'_{02} - t'_{01}, \tag{B.44}$$

where t'_{01} is the value of time on clock O watched from

M when E_1 occurs (at M_1, which coincides with P) and t_{02} is the value of time shown by the clock at O to the observer of M_2 when E_2 occurs (at the moment when P and M_2 coincide).

The problem is to compare $\Delta t_0'$ and Δt_A. According to the pre-relativistic concept, which admits a universal time, $\Delta t_0' = \Delta t_1$. According to the theory of relativity, $\Delta t_0' \neq \Delta t_A$, their relationship being known. Now it is the physical sense which is to be analyzed.

For this purpose, let us recall that it was assumed that the reference frames were inertial, and this requires the satisfying of eqns. of type (B.38), which means:

$$\tau_0 = v_0 t_0 + b_0, \ \Delta\tau_0 = c_0 \Delta t_0, \qquad \text{(B.45)}$$

$$\tau_A = c_A t_A + b_A, \ \Delta\tau_A = c_A \Delta t_A, \qquad \text{(B.45')}$$

subscript zero denoting the frame Oxyz and subscript A the frame Aαβγ. Likewise, synchronization of clocks in Oxyz yields $\Delta t_0 = t_{02} - t_{01} = \Delta t_0'$, where t_{01} and t_{02} are the instants of E_1 and E_2, respectively, both constant in O. Fig. B-2 shows the two frames Oxyz and Aαβγ in a particular case, which does not restrict the generality of observations.

The two positions of frame Aαβγ, and implicitly of point P, are simultaneous with the events E_1 and E_2.

In this case the following conclusions have to be drawn:

(a) If the relativistic postulate claiming that the speed of light is a unique constant (c) for all inertial reference frames is accepted, then, in eqns.

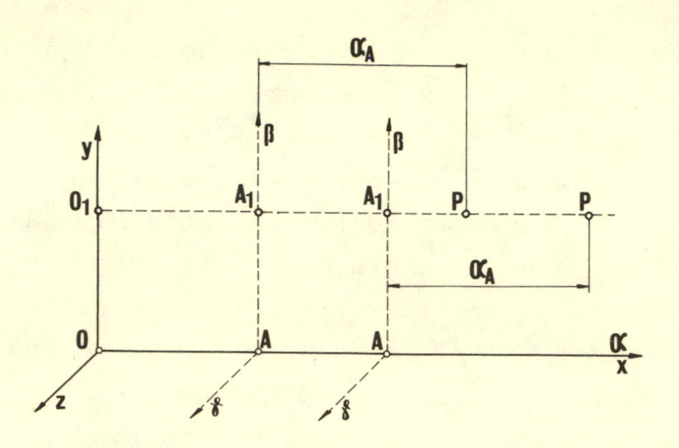

Fig. B-2

(B.45) and (B.45'), $c_0 = c_A = c$ and, because the relative motion of frames imposes $\Delta\tau_0 \neq \Delta\tau_A$, it is necessary that $\Delta t_0 \neq \Delta t_A$.

In terms of eqns. (B.45) and (B.45') only, this postulate does not appear as necessary. It seems that a suitable selection of variables t_0 and t_A (implicitly, the speeds of clocks in the two reference frames) would yield $\Delta t_0 = \Delta t_A$ and thus $c_0 \neq c_A$, as classical physics permits. This aspect requires additional analysis.

(b) If the relative motion of reference frames is inertial, i.e., equations (B.45) and (B.45') hold true, all transformation equations that relate two (or more) such reference systems are linear, which points immediately to the existence of the constant velocity of the relative motion of such frames.

(c) If various factors (in fact, gravitational and acceleration fields) alter the linear character of the dependence of the quantities involved in the phenomena admitted as timing phenomena and those related to the propagation of information, then the inertiality of motion disappears and synchronization of clocks becomes generally impossible.

It may be attempted on portions sufficiently small to be able to approximate the linearity of dependence, that is to approximate clocks employed and propagation of information (propagation of a light wave) belonging to the same equivalence class.

(d) The motion of propagation of information is necessarily associated with any observation, any measurement, that is, with the discovery and study of any transformation in the universe. It is that form of motion which accompanies any other form of motion observed, and all clocks have to be equivalent with the motion of transport of information. Otherwise, synchronization of clocks becomes impossible.

Since the variable of time, with its abstract character, resulted from the synthesis of clocks in a certain equivalence class, one may in principle _choose_ the equivalence class.

In fact, the equivalence class of the motion of the information carrier (propagation of light, more generally of an influence or perturbation) is imposed objectively.

Therefore the propagation of information is a standard motion. It must be taken as such in all inertial frames.

B.I.3.5. _The character of the unique constant velocity of light in all inertial frames_. Some of the considerations above have pointed to the need for utilizing clocks from the same equivalence class, with the motion of the information carrier considered as timing parameter. This ensures that the velocity of propagation of information (or of long range interaction), which is considered to be the speed of light, is a constant due to the linear dependence between space travelled through by information (influence, or perturbation) and the timing parameter (that is, time). However, nothing has so far shown that this constant has to be ascribed the same value in any reference frame. Obviously, only inertial frames are considered, in which synchronization of clocks is possible if one employs clocks that run all at the same rate; in such frames a clock that belongs to the same equivalence class as the propagation of information (influence) at a certain point N remains in the same class if placed at any other point P; if the clock is at rest with respect to the reference frame considered both at N and P, it shows the same time interval between events E_1 and E_2 occurring at a point M also attached to the frame.

Consider now n inertial frames $O_i x_i y_i z_i$ (i = 1, 2, ..., n). The group of transformations which achieve the transfer from one frame to the other admits the existence of some invariants.

In classical, pre-relativistic physics, time is the same in all reference frames.

Therefore, the mechanical motion of a material system with h degrees of freedom against any frame $O_i x_i y_i z_i$ obeys the same law expressed, e.g, in the holonomous case, by equations:

$$\frac{d}{dt}\left(\frac{\partial E}{\partial \dot{q}_i}\right) - \frac{\partial E}{\partial q_i} = Q_i; \qquad i = 1, 2, \ldots, h, \tag{B.46}$$

where q_i are the generalized coordinates, $\dot{q}_i$ the generalized velocities, E the kinetic energy and:

$$Q_i = \sum_{s=1}^{n} \bar{F}_s \cdot \frac{\partial \bar{r}_s}{\partial q_i} \tag{B.47}$$

is the generalized force ($\bar{F}_s$ are active forces whose work is non-zero and $\bar{r}_s$ the position vectors of their points of application).

The differential equations of mechanical motion are the same in all inertial frames.

Consider now Maxwell's equations, which describe the general laws of the electromagnetic field, equations which are the following:

$$\text{curl } \bar{H} = \bar{I}_c + \frac{\partial \bar{D}}{\partial t}, \tag{B.48}$$

$$\text{curl } \bar{E} = -\frac{\partial \bar{B}}{\partial t}, \tag{B.49}$$

$$\text{div } \bar{D} = \rho, \tag{B.50}$$

$$\text{div } \bar{B} = 0, \tag{B.51}$$

where $\bar{H}$ is the strength of the magnetic field, $\bar{I}_c$ the conduction current, $\bar{D}$ the electric induction, $\bar{E}$ the strength of the electric field, $\bar{B}$ the magnetic induction, and ρ the density of electric charge. These equations are joined by the equation of continuity $\text{div } I_c = \frac{\partial \rho}{\partial t}$.

Any electromagnetic phenomenon has to be seen (according to classical, pre-relativistic physics) by any observer attached to any of the frames $O_i x_i y_i z_i$ ($i = 1, 2, \ldots, n$) as described by the same equations

(B.48), (B.49), (B.50) and (B.51).

Thus, in inertial frames the laws of nature are expressed identically. This truth, unanimously recognized in pre-relativistic physics, which assumes time to lapse uniformly, all throughout the universe and in any frame, has ceased to be verified by experiments when their accuracy has reached a high level.

E.g., in order that Maxwell equations take the same form in all frames $O_i x_i y_i z_i$, Lorentz transformations have to be employed (which were known before Einstein's theory of relativity was formulated). However, Lorentz could not justify rigorously the transformations he advanced; justification was supplied by the theory of relativity, which in turn is based on the axiom regarding the invariance of the speed of light in all inertial frames, an axiom which seemed surprising and has remained so for many physicists, even after its backing by experiment.

In the following, this aspect, which is considered very important, is considered, because in the theory of relativity it is not so much the complications of a mathematical kind that matter (it is true they exist, though mostly in general relativity) but the interpretations and assimilation of the physical phenomenon, which represent the subtlest part.

The equations which describe the motions (transformations) in nature as, e.g., (B.46), (B.48), (B.49), (B.50) and (B.51), were established in classical physics without considering the major role the propagation of information and interaction (influence) plays in the determination by the observer of experimental data.

In fact, any observation, any value of a measured

quantity, and any variation depend both on the phenomenon observed, on the variations it implies, and on the way information gets to the observer, a factor which classical physics has not considered in its deduction of equations that describe transformations.

Therefore, one and the same phenomenon noted by two or more observers is described by the same equations, regardless of the positions of observers with respect to the site where the phenomenon occurs, of their mutual positions, of their motions with respect to the point observed, and of the relative motion, provided that these motions are not accelerated. (The accelerations introduce, according to pre-relativistic physics as well, kinematic forces).

However, because any phenomenon observed is associated with the transport of information to the observer, there are differences between:

(a) Experimental findings based on measurements carried out by an observer O and the conclusions obtained by calculation, utilizing the equations which describe the unfolding of the phenomena studied, if the observer O moves, even inertially, with respect to the domain D of space where the phenomenon takes place. (If O is at rest with respect to D and both O and D are at rest with respect to an inertial frame, the measurements of observer O coincide with calculations.)

(b) Measurements of two observers O_1 and O_2 if O_1 and O_2 are in relative motion (it may be inertial). Obviously, the statement extends to the case of n observers $O_1, O_2, \ldots, O_n$.

Therefore, the cause which determines these differences is the motion of propagation of information, motion which is composed with that of the observers (and the result of this composition depends on the observer's motion). The equations of classical physics did not predict such differences; they ignored the propagation of information (influence).

The progress of experimental techniques has allowed the detection of these differences (which are very small at usual velocities), and so has triggered the well-known crisis in physics (at the beginning of XX-th century) and has imposed the need to correct the natural approach.

There were, in principle, two possible ways of solving the problem:

(1) A standard phenomenon is chosen, recorded in a frame $O_0x_0y_0z_0$ (which becomes thus a privileged frame, the basic frame for the study of all motions, those of the other frames included). This standard phenomenon is considered the timing phenomenon which supplies the universal time. All motions (the timing ones included) with respect to $O_0x_0y_0z_0$ and another system Oxyz (moving with respect to $O_0x_0y_0z_0$) should have been expressed by equations which are corrected for the phenomenon of transport of information.

This correction would have resulted in the laws of nature being expressed differently in frames that are in relative motion, even inertial motion.

Such an approach, extrapolated from classical (pre-relativistic) scientific thinking, assumes the existence of a frame ($O_0x_0y_0z_0$) admitted as a basic system and a unique timing phenomenon for the whole universe.

(2) Another possibility of solving the problem of participation of the information (influence) propagation in observations and measurements lay in adopting a

manner of formulation that eliminates from the equations of motion (transformations) the influence of this phenomenon.

This may be achieved if one chooses as standard timing motion in any inertial frame Oxyz the motion of propagation of information (influence) taken with respect to frame Oxyz itself. Thus, all motion in Oxyz is compared to propagation of information in Oxyz as well.

Then, it is obvious that the motion of Oxyz does not matter in the writing of equations, provided it is inertial.

The time of various frames may cease to be the same when they move with respect to each other, because, due to their relative motion, propagation of information occurs in a different way in different frames and the timing motion chosen differs from one frame to the other.

The first and most important consequence of this procedure is the following: in any frame, the speed of propagation of information (speed of light) is equal to unity because its motion is referred to itself. Getting from unit to constant c (speed of light) is simply a matter of scale.

This is the origin of the invariance of the speed of light.

It follows that when Einstein formulated the axiom of invariance in all inertial frames, he implicitly admitted propagation of information as a timing motion.

Therefore it should be observed that:

(1) The invariance of the speed of light loses its axiomatic character. Interpretation of the concept of

time becomes the factor that generates relativistic physics.

(2) All timing motions represented by propagation of information in various reference frames belong to the same equivalence class if the frames in question are all inertial (because the transformation relations are linear).

(3) In eqns. (B.45) and (B.45') one gets $c_0 = c_A = c$ and $t_0 \neq t_A$.

(4) In general, in the case of non-inertial frames, synchronization of equivalent clocks is not possible.*

B.I.3.6. Relativity of time and the problem of several information carriers. Once the speed of light is accepted as a universal constant, the relativity of time, as it appears in the theory of relativity, follows through reasoning and calculations known and presented in the first part of this book.

The theory of relativity is structured around the hypothesis that the speed of light could not be exceeded.

No experiment has contradicted this hypothesis before the birth and crystallization of the theory of relativity.

For the sake of rigour, it would be more suitable to consider the set of all observers (be it finite or infinite) consisting of the whole of some equivalence class. All observers in one and the same equivalence class are sensitive to a certain information carrier. (By "information carrier" is understood the form of

---

* See §§ C.I.2.2., C.II.2.2., and C.III.2.2.

energy whose propagation carries information from the universe at highest speed, higher than that of other forms of energy which may influence the observer). Otherwise the phenomena studied by a particular observer assume a particular information carrier.

Consider Ω_0 the set of all observers and K_1, K_2, ..., K_n the equivalence classes that Ω_0 is divided into.

The overall number of classes may be finite or infinite.

K_i is a set of observers $Q_{i\alpha}$, with $\alpha = 1, 2, \ldots, p_i$. Obviously, p_i represents the number of observers in class K_i and may also be finite or infinite.

With $i \neq j$, it follows that

$$K_i \cap K_j = \emptyset. \tag{B.52}$$

Likewise, it is easy to see that:

$$\bigcup_{i=1}^{n} K_i = \Omega_0. \tag{B.53}$$

Consider now E, an event at a point M of space, and τ_i $(i = 1, 2, \ldots, n)$, the space travelled through by an information carrier A_i.

The observers $O_{1\alpha}$ perceive A_1, observers $O_{2\alpha}$ A_2 etc.

Then $A_1, A_2, \ldots, A_n$ may not be directly compared, because no observer perceives both A_i and A_j if $i \neq j$.

For each observer $O_{i\alpha}$ the state quantities q_{ij} may be expressed by functions

$$q_{js} = q_{js}\,(t_i), \tag{B.54}$$

whatever the system Γ_s whose state is defined by q_{js} ($j = 1, 2, \ldots, \gamma_s$). Obviously, t_i is the time of observers $O_i \in K_i \subset \Omega_0$ and thus, according to the specifications above, τ_i being the distance covered by the information carrier (in fact light, but in this reasoning the case when τ_i represents the space travelled through by light is a particular one), one can write:

$$\tau_i = c_i t_i + b_i. \tag{B.55}$$

This is so because all of the observers O_{is}'s clocks are necessarily equivalent with the motion of transport of information, which confers on the relationship between τ_i and time (a result of the synthesis of all timing parameters, a variable which is a measure of universal motion) a linear nature.

Once the linear nature is established, the proportionality factor, which multiplies time, c (speed of information propagation), has to be the same for all frames to which motions are referred, i.e., it is awarded the feature of invariance (as was done with the speed of light, whatever the information carrier). Obviously, the statement refers to observers $O_{i\alpha}$ in class K_i.

For any observer $O_{i\alpha} \in K_i \subset \Omega_0$ ($\alpha = 1, 2, \ldots, p_i$; $i = 1, 2, \ldots, n$) the theory of relativity remains adequate, with all its relations and arguments provided that there is a constant c_i which plays the role of the speed of light, for each class K_i.

Obviously, $i \neq j$ implies $c_i \neq c_j$.

Consider now a system formed of n observers, each belonging to one class K_i. The subscript may be discarded because there is just one for every class K_i. Thus, consider a system S_0 formed of n observers $O_i \in K_i$ $(i = 1, 2, \ldots, n)$, all at rest with respect to each other.

S_0 occupies a domain D_0 in the space of the reference frame Oxyz (with respect to which S_0 is at rest). The dimensions of D_0 are negligible with respect to the distances from which the information regarding the events noted by observers $O_i \in K_i$ travels.

The interesting problem arises now as to the consequences of the possibility that observers O_i communicate with each other for extremely short times.

Naturally, it was assumed that the observers in two distinct classes, $O_i \in K_i$ and $O_j \in K_j$ with $i \neq j$, cannot communicate, because, if they could, the following dependence was established:

$$\tau_i = f\ (\tau_j), \tag{B.56}$$

and, as a consequence, if $\frac{d\tau_i}{d\tau_j} > 1$, the class K_j would no longer be distinct but would be assimilated (included) in K_i, whilst if $\frac{d\tau_i}{d\tau_j} < 1$, K_i would disappear as a distinct equivalence class, becoming a part of K_j. Both cases are contrary to the initial hypothesis, that K_i and K_j $(i \neq j)$ are distinct, defined as above.

As mentioned above, it would be interesting to study what happens when observers belonging to the same equivalence class may communicate instantaneously at large intervals (eventually at random).

In other words, we study the case of observers from different equivalence classes, isolated nearly all the time but not permanently. This amounts to admitting that, in the case when the observers communicate only during sufficiently short time intervals, which may be considered zero, they may belong to different classes.

The following possible cases are considered:

(1) The system S_0 consisting of the n observers $O_i \in K_i$ is at rest with respect to points M_β where the events observed, E_B, occur. That is, M_β are the fixed points of Oxyz.

In this case the analysis needed is very simple.

Consider events E_β and E'_β, both occurring in M_β. According to the way that the sets of observers were constructed, by virtue of the criterion which separates and defines their equivalence class, if an observer $O_i \in K_i$ notes events E_β and E'_β, at times $t_{i\beta}$ and $t'_{i\beta}$ with $t'_{i\beta} > t_{i\beta}$, and another observer $O_j \in K_j$ notes (following instructions) the same events at times $t_{j\beta}$ and $t'_{j\beta}$ with $t'_{j\beta} > t_{j\beta}$ (obviously both O_i and O_j belong to S_0), then usually one can not establish any relationship between $t_{i\beta}$ and $t_{j\beta}$, $t'_{i\beta}$ and $t'_{j\beta}$ and consequently between $\Delta t_{i\beta} = t'_{i\beta} - t_{i\beta}$ and $\Delta t_{j\beta} = t'_{j\beta} - t_{j\beta}$. This statement seems obvious, taking into account that the observers are isolated from each other, that is do not communicate. (If they communicated, the source of information of one of them would be that of others as well, and they would not longer belong to different classes.)

It is assumed that between observers O_i and O_j a very short quasi-instantaneous communication is established.

The time of communication is t_{ci} for O_i and t_{0j} for O_j.

When $t_{ci} > t_{i\beta}$ and $t_{0j} < t_{j\beta}$, O_j may find out about event E_β before the information is received through the information carrier. For the observer O_i the event E_β appears under these conditions as lagging by $\delta t_j = t_{j\beta} - t_{cj}$.

It is assumed now that the communication between O_i and O_j is resumed, and O_i notes this second communication at time t'_{ci} and O_j at t'_{0j}.

It is further assumed that:

(a) $t'_{ci} > t'_{i\beta}$ and $t'_{cj} > t'_{j\beta}$. If O_j learns about E'_β through O_i (when O_i is at time t'_{cj}), this happens after the information about E'_β has reached the observer. Thus, for O_i the interval between E_β and E'_β is $\Delta t_{i\beta} = t'_{i\beta} - t_{i\beta}$ and for O_j the interval between the same events is $\Delta t_{j\beta} = t'_{j\beta} - t_{cj}$, considering that the interval is delimited by the arrival of the first information about E_β and E'_β.

(b) $t'_{ci} > t'_{i\beta}$ and $t'_{cj} < t'_{j\beta}$. If O_i communicates with O_j about the event E'_β, the interval between events for O_i is $\Delta t_{i\beta} = t'_{i\beta} - t_{i\beta}$ and for O_j the time interval separating E'_β and E_β becomes $\Delta t_{j\beta} = t'_{cj} - t_{cj}$.

(c) The same expression is valid for the interval $\Delta t_{j\beta}$ when $t_{ci} > t_{i\beta}$; $t'_{ci} > t'_{i\beta}$ and $t_{cj} < t'_{cj} < t_{j\beta} < t'_{j\beta}$. It is just that, in such a case, O_j perceives both events E_β and E'_β twice, first through O_i and then through O_j's own information carrier.

It is easy to see that in previous cases there also exists a double perception by O_j of one of events E_β and E'_β or both, due to the two sources of information: O_i and O_j's own information carrier, a phenomenon which is equivalent to outrunning, in the case of some observers, their own time lapse.

It is important to mention that the moment when the event occurs for any observer is the arrival of the first information regarding the event.

Regarding the relation between $\Delta t_{i\beta}$ and $\Delta t_{j\beta}$, this may only be determined by imposing various conditions and making assumptions of an artificial character.

It is interesting to follow now what happens with a set Ω of observers as yet undivided in disjoint equivalence classes (according to the criterion above) but made up of subsets which distinguish themselves from one another as follows:

Assume the set μ_i formed of observers $O_{i\gamma_i}$ (γ_i = 1, 2, ..., q_i) whose information carrier is A_i and the subset μ_{i+1} formed of observers $O_{(i+1)\gamma_{i+1}}$ (γ_{i+1} = 1, 2, ..., q_{i+1}) whose information carrier is A_{i+1}. All observers in μ_{i+1} may also distinguish the form of energy the agent A_i carries. However A_{i+1} is more rapid. That is, if t_i is the time for observers $O_{i\gamma_i}$ and t_{i+1} is the time for observers $O_{(i+1)\gamma_{i+1}}$, the time being defined as usual by a general synthesis of all transformations noted by the observers, then the space τ_i travelled through by the agent which transports information to $O_{i\mu_i}$ is

$$\tau_i = c_i t_i + b_i, \qquad \text{(B.57)}$$

and, for the observers in μ_{i+1}, one can write:

$$\tau_{i+1} = c_{i+1} t_{i+1} + b_{i+1}, \qquad \text{(B.57')}$$

with obvious notations.

Because $O_{(i+1)\gamma_{i+1}}$ perceives both A_i and A_{i+1}, these observers may always establish a dependence of the type:

$$\tau_i = f(\tau_{i+1}). \qquad \text{(B.58)}$$

The fact that A_{i+1} is a faster agent than A_i is expressed by the equation:

$$\frac{d\tau_i}{d\tau_{i+1}} = \frac{d}{d\tau_{i+1}} f(\tau_{i+1}) < 1. \qquad \text{(B.59)}$$

Obviously, taking into account that τ_i and τ_{i+1} are spaces, $\frac{d\tau_i}{d\tau_{i+1}}$ is a dimensionless quantity. Under these circumstances, $i = 1, 2, \ldots, n-1$, n being the total number of sets μ_i defined as shown.

Taking two such sets μ_j and μ_i, if $j > i$ (that is A_j is more rapid than A_i) the set μ_j is said to be $j-i$ times higher than the set μ_i. It is specified that, in general, the observers of μ_i do not communicate with those in μ_j and do not perceive A_j.

If there exists the possibility that, in certain circumstances and special conditions, an observer of μ_j communicates an event or sequence of events to an observer of μ_i $(i < j)$, then the observer in μ_i <u>may outrun time</u> with respect to the other observers from μ_i, and this will be noted by all observers in classes μ_k $(k \geqslant i)$. This is so if one admits that communication is normally possible from observer μ_i to μ_k, $i \leqslant k$, but reverse communication is an exception which takes place at certain moments, dictated by the simultaneous fulfilment of an ensemble of conditions.

If the distances between all $\sum_{k=1}^{n} q_k$ observers are sufficiently small to be neglected, then $O_{k\gamma_k} \in \mu_k$ notes that, by communication with $O_{j\gamma_j} \in \mu_j$, $O_{i\gamma_i} \in \mu_i$ has outrun the time in a way whose quantitative expression may be readily established, as in the case of sets K_i of observers.

Meanwhile, if events E_β and E'_β are recorded by $O_{i\gamma_i}$ at t_i and t'_i and by $O_{j\gamma_j}$ at t_j and t'_j, then $O_{j\gamma_j}$ notes event P_β, which consists of the recording by $O_{i\gamma_i}$ of E_β at t_{ij}, and P'_β consisting of the recording by $O_{i\gamma_i}$ of E'_β at t'_{ji}.

Even though $t'_j - t_j = t'_{ji} - t_{ji}$, O_j notes that $t'_{ji} > t'_j$ and $t_{ji} > t_j$.

Obviously t_k is the time for observers of μ_k ($k = 1, 2, \ldots, n$).

In other words, under the known conditions, observers μ_j note that observations of observers μ_i ($i < j$) lag, which is due to the fact that agent A_j is faster than A_i. In fact these findings may underlie the comparison of information carriers in order to establish their hierarchy and their relative velocity.

In fact, the case is similar to that encountered in the set K_i of observers except that, considering now the sets μ_i, there may exist the possibility of experimental findings by observers in the higher sets upon those in the lower sets.

It was obviously assumed that all observers in the sets μ_k are at rest with respect to a frame Oxyz and so are in the same frame as the points M_β where events E_β and E'_β occur, as in the case of the observers that formed the sets K_i.

(2) The system of observers is at rest with respect to the reference frame Oxyz; M_β, a point where events E_β and E'_β occur, moves with respect to Oxyz.

Consider Δt_β, the proper time interval, delimited by events E_β and E'_β and measured by a clock attached to M_β.

The velocity of M_β relative to Oxyz during the time between E_β and E'_β being v_i, an observer (bound to

Oxyz) who receives information from any carrying agent notes a time interval:

$$\Delta t_i = \frac{\Delta t_\beta}{\sqrt{1 - \frac{v_i^2}{c_i^2}}}, \tag{B.60}$$

which is an alternative to eqn. (A.18) (with c_i the speed of the agent carrying information to the observer considered); that is, the observer measures and orders the transformations with a variable t_i such that $\tau_i = c_i t_i + b_i$ (τ_i being the space covered by the information carrier, a parameter which changes as the information propagates) and v_i is the velocity of M_β calculated with time t_i.

In particular, as is the case with eqn. (A.94), the information carrier may be light, and in that case c_i is the speed of light.

In the case of some observers belonging to different equivalence classes K_i or to some different sets μ_i, the Δt_i will differ.

In other words, it is obvious that:

$$\Delta t_i = \Delta t_\beta \ (1 - v_i^2 c_i^{-2})^{-\frac{1}{2}}$$

and

$$\Delta t_j = \Delta t_\beta \ (1 - v_j^2 c_j^{-2})^{-\frac{1}{2}}$$

imply

$$\Delta t_i \neq \Delta t_j \text{ if } v_i^2 c_i^{-2} \neq v_j^2 c_j^{-2}.$$

This fact may be established or verified by experiment only if the observers who determine Δt_i and Δt_j between the same events E_β and E'_β communicate at least in one sense, i.e., at least one of the observers, not any one of them, but the one in the higher class, may follow the processes experienced by the other, even though this communication goes only in one sense. In the case when communication between observers is reciprocal and permanent, neither their findings nor their time can differ any longer, regardless of the fact that one observer gets the information indirectly, through the other.

Thus communication of observers is either instantaneous, i.e., on a set of intervals of quasi-zero measure or in a unique sense, in which case only the observer belonging to a higher set μ_j is able to note the processes occurring with an observer that belongs to μ_i $(i < j)$.

The communications called instantaneous may enjoy reciprocity and lead to changes of time lapse for one of the observers.

These considerations may be extended by building a theory in which observers may transfer from one class K_i into another one, K_j, or from a set μ_i to a set μ_j $(i \neq j)$, but <u>any observer is at any time in one single class K_i (or K_j) or one single set μ_i (or μ_j)</u>. The time of such an observer would undergo either outruns or delays.

<u>Note</u>. The truths of the theory of relativity may be checked by experiment for any observer, whatever the information carrier, provided that c, the speed of light is replaced by c_i, the speed of <u>that carrier</u>.

Naturally, in case of our nature and of the set of

observers one can communicate with, the carrier is light (the electromagnetic wave).

At the same time, unusual communication (perhaps a random, exceptional phenomenon, which is theoretically possible) between the observers employing various motions of propagation of information will bring about variations, real disturbances, in the time lapse of some observers (in the lower classes).

If science supplies evidence for the existence of parapsychological phenomena, now the object of controversy, their rigorous explanation could be attempted in terms of these considerations by admitting that there may exist motions faster than light which are perceived only in exceptional instances.

B.II. EXAMPLES

In order to substantiate some points of view expressed above, some examples are given next.

B.II.1. The irreversibility and continuity of time

B.II.1.1. The probabilistic irreversibility. Consider an observer O obliged to confine his research to a universe consisting of n material systems S_i (i = 1, 2, ..., n), each characterized by h_i state variables q_{ij} (j = 1, 2, ..., h_i) whose variations are mutually independent. It is assumed that the variation of quantities q_{ij} is quantized, so that $q_{ij} = \lambda_{ij} \nu_{ij}$, with λ_{ij} a natural number and ν_{ij} the quantum of q_{ij}. To make things simpler, suppose that the variations of q_{ij} (for any i or j) are stepwise, i.e., no variation implies a jump larger than one quantum. However this value of the jump may be positive, negative or zero.

Thus, the next value of $q_{ij} = \lambda_{ij} \nu_{ij}$ is

$$q_{ij} = (\lambda_{ij} - 1) \nu_{ij},$$

$$q_{ij} = (\lambda_{ij} + 1) \nu_{ij},$$

or else it remains

$$q_{ij} = \lambda_{ij} \nu_{ij}.$$

Admitting that material systems S_i are subjected to mutual and external influences and that, as a result of these influences, their states, expressed by q_{ij} values, change, let $q_{ij} = (\lambda_{ij} - 1) \nu_{ij}$, with $i = 1, 2, \ldots, n$, $j = 1, 2, \ldots, h'_i$; $q_{ij} = \lambda_{ij} \nu_{ij}$, with $i = 1, 2, \ldots, n$, $j = h'_i + 1, h'_i + 2, \ldots, h''_i$ ($\geqslant h_i$); $q_{ij} = (\lambda_{ij} + 1) \nu_{ij}$, with $i = 1, 2, \ldots, n$, $j = h''_i + 1, h''_i + 2, \ldots, h_i$ (and $h''_i \leqslant h_i$) be the values which characterize the state of the ensemble of systems S_i immediately before the state $q_{ij} = \lambda_{ij} \nu_{ij}$. It is possible that $h'_i = 0$, $h''_i = h'_i$ or $h''_i = h_i$. The ordering of states, i.e., the lapse of time, is realized by the observer by means of the variation of one or more of the variables q_{ij}; e.g., to fix ideas, assume it is q_{11}. This means that $q_{11} = \lambda_{11} \nu_{11}$ and λ_{11} represents the total number of states of the whole microworld $S = \bigcup_{i=1}^{n} S_i$ which is the object of O's observations.

In the microworld S, q_{11} fulfils the conditions required of a timing parameter. If S's state, characterized by $q_{ji} = \lambda_{ij} \nu_{ij}$, corresponds to a value

$q_{11} = \lambda_{11} \nu_{11}$, then the immediately anterior state corresponds to a value $q_{11} = (\lambda_{11} - 1) \nu_{11}$ and the one following immediately to $q_{11} = (\lambda_{11} + 1) \nu_{11}$. Therefore, λ_{11} points to a state of the microworld S. The existence of the variation of q_{11} alone occuring in only one sense (i.e., the set of q_{11} values is ordered, this being one of the reasons for selection of q_{11} as timing parameter) would be sufficient to confer on time the character of irreversibility.

The following hypothesis is accepted:

(a) All other variations q_i are reversible and random because there exists for any q_{ij}: the probability $P_{ij}^{(0)}$ that in state $\lambda_{11} + 1$ (following after λ_{11}) it remains with the value unchanged, $q_{ij} = \lambda_{ij} \nu_{ij}$, the probability $P_{ij}^{(+1)}$ that in state $\lambda_{11} + 1$ the value is $q_{ij} = (\lambda_{ij} + 1) \nu_{ij}$, and the probability $P_{ij}^{(-1)}$ that state $\lambda_{11} + 1$ assumes $q_{ij} = (\lambda_{ij} - 1) \nu_{ij}$.

At the same time, $P_{ij}^{(0)} + P_{ij}^{(-1)} + P_{ij}^{(+1)} = 1$.

(b) Whenever deciding on the irreversibility (or reversibility) of time, the observer O ignores the variation of the parameter q_{11}.

(c) The observer O ignores the phenomenon of transport of information; the error that originates in this is smaller, the smaller the extent of the microworld $S = \bigcup_{i=1}^{n} S_i$.

(d) For the sake of simplicity, S_1 is characterized only by q_{11} ($h_{11} = 1$).

The problem now arises to establish the probability (denoted by P) that the state $\lambda_{11} + 1$ of the microworld $S' = \bigcup_{i=2}^{n} S_i$ is identical with the state $\lambda_{11} - 1$, leaving aside q_{11}, which varies irreversibly and rules out the possibility of reversibility of S' motion i.e., of the identity of $\lambda_{11} - 1$ and $\lambda_{11} + 1$ states.

With the notations above $P_{11}^{(0)} = P_{11}^{(-1)} = 0$ and $P_{11}^{(+1)} = 1$.

According to the specifications and some elementary rules of the theory of probability, the probability P sought is given by:

$$P = \prod_{i=2}^{n} \left[\prod_{j=1}^{h_i'} P_{ij}^{(-1)} \prod_{j=h_i'+1}^{h_i} P_{ij}^{(0)} \prod_{j=h_i''+1}^{h_i} P_{ij}^{(+1)} \right]. \qquad \text{(B.61)}$$

Obviously $0 \leqslant P_{ij}^{(\alpha)} \leqslant 1$, whatever $i = 2, 3, \ldots, n$; $j = 1, 2, 3, \ldots, h_i$; $\alpha = -1, 0, +1$.

It follows that, for n and h_i finite and natural numbers, $P > 0$ (and naturally $P < 1$); thus the observer O could note a reversibility of states in his own universe, time included. However this is less probable for higher n, and:

$$\lim_{n \to \infty} P = \lim_{n \to \infty} \prod_{i=2}^{n} \left(\prod_{j=1}^{h_i'} P_{ij}^{(-1)} \prod_{j=h_i'+1}^{h_i''} P_{ij}^{(0)} \prod_{j=h_i''+1}^{h_i} P_{ij}^{(+1)} \right) = 0; \qquad \text{(B.62)}$$

likewise:

$$\lim_{h_i' \to \infty} P = \lim_{h_i'' \to \infty} P = \lim_{h_i \to \infty} P = 0, \qquad \text{(B.63)}$$

whatever $i = 2, 3, \ldots, n$, which is obvious and together with (B.62) attests to the <u>probabilistic irreversibility</u> of states in a complex universe and implicitly the probabilistic irreversibility of time.

Eqn. (B.61) also shows that any $P_{ij}^{(\alpha)} = 0$, for some $i = 2, 3, \ldots, n$; $j = 1, 2, 3, \ldots, h_i$; $\alpha = -1, 0, +1$, implies $P = 0$. That is, if at least one transformation is irreversible for whatever n and h_i, then no one state of the whole universe is repeated, and therefore time is irreversible.

B.II.1.2. <u>The continuity of the time variable</u>. Consider the random variable n which takes values from the set $\lambda_{ij}\, \nu_{ij}$ ($i = 2, 3, \ldots, n$; $j = 1, 2, \ldots, h_i$).

Taking into account that $\lambda_{ij}\, \nu_{ij} = q_{ij}$ are state quantities, they may be expressed in different units. Starting from this fact, one chooses a system of units, that is, chooses numbers q_{ij}. These numbers are the values of the random variable u.

Thus u appears as taking the values q_{ij} themselves, but, unlike q_{ij}, the values of u are considered dimensionless.

Consider now this set of values:

$$\{U_k\} = \bigcup_{i=2}^{n} \left\{ \bigcup_{j=1}^{h_i} \lambda_{ij}\, \nu_{ij} \right\}, \tag{B.64}$$

where the λ_{ij} are integers.

Let α_{ij} be the lowest value $\lambda_{ij}\, \nu_{ij}$ and $\beta_{ij} = \mu_{ij}\, \nu_{ij}$ the highest value $\lambda_{ij}\, \nu_{ij}$.

Then:

$$\{U_k\} = \bigcup_{i=2}^{n} \left\{ \bigcup_{j=1}^{h_i} \{\alpha_{ij}, \alpha_{ij} + \nu_{ij}, \alpha_{ij} + 2\nu_{ij}, \ldots, \alpha_{ij} + \mu_{ij}\, \nu_{ij}\} \right\} \tag{B.65}$$

where μ_{ij} is the highest number λ_{ij}.

We are now interested in determining the probability that, for at least one value u_k, $u_k \in (a, b)$, for some numbers k, a, b. This probability is denoted by P_{ab} (obviously $k = 1, 2, \ldots, \sum_{i=2}^{n} \sum_{j=1}^{h_i} j$). The probability P_{ab} is needed in some reasoning and calculations that follow next.

Let $P'_{ab} = 1 - P_{ab}$ be the probability that, for an interval [a,b], $u_k \notin [a,b]$ for $k = 1, 2, \ldots, \sum_{i=2}^{n} \sum_{j=1}^{h_i} j$.

On the [a,b] interval, the only condition imposed is that $0 < \varepsilon = b - a < |\nu_m|$, where $|\nu_m|$ is the smallest value from the set of $|\nu_{ij}|$ values and $|\nu_{ij}|$ stands for the (obviously real) number which expresses the quantum ν_{ij} after the system of units has been selected. This manner of looking at things is dictated by the need for comparing quantities (quanta ν_{ij}) which present qualitative differences and therefore are expressed by different units (but are part of the same system of units).

From the way of defining the probability P_{ab}, it follows that:

$$P'_{ab} = \prod_{i=2}^{n} \left(\prod_{j=1}^{h_i} \frac{|\nu_{ij}| - \varepsilon}{|\nu_{ij}|} \right). \qquad \text{(B.66)}$$

For suitable small (always positive) $\varepsilon = b - a$ it is seen that:

$$\frac{|\nu_{ij}| - \varepsilon}{|\nu_{ij}|} < 1$$

and as a consequence:

$$\lim_{n \to \infty} P'_{ab} = 0, \quad \lim_{h_i \to \infty} P'_{ab} = 0, \qquad (B.67)$$

for any $i = 2, 3, \ldots, n$.

And, because $P_{ab} = 1 - P'_{ab}$, it follows that:

$$\lim_{n \to \infty} P_{ab} = 1, \quad \lim_{h_i \to \infty} P_{ab} = 1, \qquad (B.68)$$

for $i = 2, 3, \ldots, n$.

In conclusion, if the number of systems observed by O tends to infinity or if (without its being necessary that the number of state variables tends to infinity) the number of state variables of a system is infinite over any interval $[a, b]$, there is at least a value u_k, no matter how small is the measure $\varepsilon = b - a$ of the interval considered. However, the interval $[a, b]$ may at any time be divided into a series of subintervals $[a_s, b_s]$, with $s = 1, 2, \ldots, \infty$, so that:

$$\bigcap_{s=1}^{\infty} (a_s, b_s) = \emptyset,$$

$$\bigcup_{s=1}^{\infty} [a_s, b_s] = [a, b], \qquad (B.69)$$

where $\emptyset$ is the empty set.

Likewise, any of these intervals ($[a_s, b_s]$ with $s = 1, 2, \ldots, \infty$) may be associated with the reasoning and calculations above.

Thus, on each interval $[a, b]$ there is not one but an <u>infinity of values u_k, however small its measure</u> $\varepsilon = b - a$, if $n \to \infty$ or $h_i \to \infty$ (for any $i = 2, 3, \ldots, n$).

Thus, one deduces that for a finite number of systems observed ($n \to \infty$), or if one of these contains an infinity of state quantities ($h_i \to \infty$), a <u>continuous</u> set of values is (probabilistically) imposed for u_k. Let u (without subscript k) denote the variable that takes its values from this set.

However, any transformation in the observer O's universe in general assumes the variation of at least one quantity q_i, which, on the grounds of the foregoing, means the variation u, on the one hand, and the variation of q_{11}, which bears some timing nature, on the other hand.

Consider a state A of observer O's universe, corresponding to a value $q_{11}^{(a)}$ of the timing parameter, and another state B of the same universe, corresponding to $q_{11}^{(b)}$ of the same parameter.

It is assumed that $q_{11}^{(b)} > q_{11}^{(a)}$. Let $q_{ij}^{(a)}$ and $q_{ij}^{(b)}$ denote the values of state variables in A and B, respectively. Taking into account eqns. (B.64), (B.65) and (B.68) together with the subsequent considerations, the monotonic increasing variation of the q_{11} parameter and the fact that any variation of q_{11} implies the variation of at least one q_{ij} (and the variation of any q_{ij} implies that of q_{11}), it follows that any value $q_{11} \in [q_{11}^{(a)}, q_{11}^{(b)}]$ is an accumulation point of the set of numbers on the $[q_{11}^{(a)}, q_{11}^{(b)}]$ interval, that is, q_{11} takes values on a set of the power of the continuum.

It follows that, if n or h_i, for i = 2, 3, ..., n, and, more so, if n and h_i, tend to infinity, the probabilistic considerations above require that the timing parameter has to vary continuously (the ν_{11} quantum of its variation has to tend to zero).

Thus the continuity of time is a fact (q_{11} =

$c_{11}t + b_{11}$, with c_{11} and b_{11} constants and t the observer O's time), even though none of q_{ij} state variables varies continuously, but their number is infinite. If at least one of the quantities q_{ij} varies continuously (is not quantized), then the variation of the timing parameter becomes necessarily continuous and the reasoning and calculations above are no longer needed.

In this latter case, the continuity of the variation of q_{11} may be imposed even if n and h_i are finite for i = 2, 3, ..., n.

If the parameter q_{11} is, in particular, the angle between a mobile diameter of one circle and a fixed diameter, then one can write $q_{11} = 2k\pi + \phi$ and, instead of q_{11}, ϕ is employed whose variation if periodic. However, ϕ is associated with the number k, which also indicates the number of periods completed from the beginning of the study.

The time t of such an observer is given by the equation

$$q_{11} = c_{11}t + b_{11} = 2k\pi + \phi,$$
$$t = \frac{1}{c_{11}}[2k\pi + \phi - b_{11}], \tag{B.70}$$

for q_{11} and ϕ continuous variables; t is obviously a continuous variable. Although ϕ is cyclic, q_{11} and t's variation are monotonic increasing.

If the observer O also detects the phenomenon of propagation of information, q_{11} has to belong to the same equivalence class as that of the motion of the agent which carries information.

B.II.2. The forced open curves in Minkowski space

Consider two frames $O_1x_1y_1z_1$ and Oxyz in an inertial motion of velocity v_0 along the positive sense of superposed axes O_1x_1 and Ox. It is assumed that $O_1x_1y_1z_1$ is fixed and Oxyz is mobile.

A particle M moves on the circle $x_1^2 + y_1^2 = R^2$, $z_1 = O$ at a constant speed $v_1 = R\omega = \text{const.}$

The trajectory of this particle with respect to the frame Oxyz is the ellipse $y^2 + \frac{(x + v_0 t)^2}{1 - \frac{v_0^2}{c_2}} = R^2$, $z = O$, which is obviously mobile.

Thus, in both cases the trajectories of particle M are closed curves, either relative to $O_1x_1y_1z_1$ or to Oxyz.

In Minkowski space the trajectory is an open curve. Indeed let E_A denote the event of the transit of M through A (x_{1A}, y_{1A}, O) at time t_{1A} and E'_A the next transit of M through the same A at time $t'_{1A} = t_{1A} + \frac{2\pi}{\omega}$. Obviously in Minkowski space points E_A and E'_A do not coincide because the coordinates differ: $t_{1A} \neq t'_{1A} = t_1 + \frac{2\pi}{\omega}$.

These aspects are much too elementary to delay our attention further.

This second part of our work has intended to present some concepts regarding time; however the possibility of developing and substantiating the point of view exposed is selfevident. The examples supplied were deliberately very simple: they were intended to illustrate previous statements whose support seemed to require examples. It is possible that other introductions of the topic also needed examples (e.g., those in § B.I.3.6.) to make them clearer and more concrete. The pre-established extent of the book has not always allowed detailed presentation and numerous examples; however, the author hopes his point has been made sufficiently clear.

C. INERTIA

C.I. USE OF TIME IN DEFINING SOME ELEMENTS OF SPACE

The study of nature in general and of its particular aspects that are dealt with in this chapter requires rigorous definition of the notions of straight line and distance. An attempt is made to give definitions which are, according to the view adopted above, more suitable than those presently employed in specialized literature.

C.I.1. Definition of the straight line

Although the straight line may seem a primary, elementary notion that requires no definition, it includes subtleties that modern science has emphasized, and therefore its approach has to be rigorous.

C.I.1.1. Inertial spaces. The notions of synchronized clocks and inertial frames have been discussed above, but, for the sake of a unified presentation, the discussion is resumed briefly and supplemented.

Let there be a set (that theoretically may also be assumed infinite) of clocks that may be considered identical (except for some minor errors). If all are placed (successively or, if possible, simultaneously) in the vicinity of the same point M, they show the same rate of time lapse to an observer placed at M.

Then, one of these clocks is moved to point O, selected as origin of the space, the other ones being placed at different points P of space. In principle, at any point P of space one may place such a clock.

If the origin of these clocks ($t = 0$) are so arranged that, given the transmission of information, any observer at point P of space notes any event E occurring at origin O at the same time t, no matter which clock is considered, i.e., the clock at P or the clock at origin O, then the clocks are said to be synchronized.

This amounts to saying that any of the synchronized clocks is triggered when it receives the information that the clock at the origin O has started. Obviously this information (as any other information) is brought by a carrier (i.e., an electromagnetic wave).

Consider now two points P_1 and P_2 of the space, each provided with one clock and one observer. Also let E be an event occurring at the origin O; due to synchronization, both the observer at P_1 and P_2 note the event at the same moment t, regardless of what clock they use; their own clocks or the clock at the origin O.

However, if the observer at P_1 uses the clock at P_2 as well (besides that at P_1 or O), the event E is recorded at a time t_{12}. Likewise, if the observer at P_2 uses the clock at P_1, the event occurs at t_{21}.

It is assumed that:

(a) The clocks may be synchronized (as specified above) whatever the points P where they are placed;

(b) For any pair of points P_1 and P_2, one may write

$$t_{12} = t + \alpha_{12},\ t_{21} = t + \alpha_{21}, \qquad (C.1)$$

where α_{12} and α_{21} are constants for one and the same pair of points P_1, P_2 (i.e., for each pair P_1, P_2 there are two constants: α_{12} and α_{21}), whatever the time t when the event E occurs.

If all these conditions are fulfilled, the points P form an inertial space. The points of such an inertial space are at relative rest.

If $\alpha_{12} = 0$, the points P_1, P_2 and O are said to be colinear. In this case $\alpha_{12} < 0$ and $l_{21} = -\frac{1}{2}\alpha_{21}c$ is the distance between P_2 and P_1; c stands for the speed of the information carrier (which is known in nature to be the speed of light). One can abandon the abstract variable of time and refer the general motion to the speed of information transmission, in which case it turns out that c = 1. Then eqns. (C.1) should be suitably rewritten. (This aspect of referring directly to the primary phenomenon was analyzed in Chapter B of this book.)

C.I.1.2. The straight line. Let P_0 be a point in an inertial space S whose origin is O, and suppose an event E occurs at O. All observers at points $P \in E$, $P_0 \in S$ included as well as the observer at O, note the event E at the same time t if they use their own (synchronized) clocks or the clock at O. Obviously observer's own clock in P is that placed at point P. However the observer at $P_0 \in S$ does not generally note the event E at time t if use is made to measure the time of the clock at point $P \in S$ $(P \neq O)$. In this case he associates with the event

E a time $t_{P_0P} = t + \alpha$, where α is a constant that depends on the pair of points P_0 and P.

The locus of all points $P \in S$ whose $\alpha = 0$, and therefore $t_{P_0P} = 1$, is the <u>line segment</u> OP_0 of space S.

The locus of all points in space S where, when one places P_0 there the property stated above stands, is the <u>straight line</u> determined by O and P_0 in space S.

<u>Notes</u>

(1) This definition is similar to the one already known that considers the straight line to coincide with the light beam; however it offers the advantage that defining the light beam or dealing with the photon trace is no longer required.

(2) For any given pair of points $P \in S$ and $P_0 \in S$, the manner of defining the straight line given above is valid. Indeed, one chooses P as origin of space S (with all that implies), and one defines the straight line determined by P_0 and P according to the procedure described above for P_0 and O.

(3) If, in the inertial space S of origin O, the observer at O employs the clock at P to determine the moment when the event (at O as well) occurs, a time $t_{OP} = t + \alpha_{OP}$ is found, when t is the time supplied by the clock at O and $\alpha_{OP} < 0$, and $l_{OP} = -\frac{1}{2}\alpha_{OP}\, c$ is the distance between O and P (c stands for the speed of light).

This statement is fully justifiable since, as was specified, the clocks in space S are synchronized. This argument also holds for $l_{21} = -\frac{1}{2}\alpha_{21}\, c$, the distance between points P_1 and P_2 colinear with O.

If one has to express the distance between any two points, the same procedure is applied, one of the points being taken as pole of the inertial space S.

(4) Whatever was specified and defined is based, as one easily notices, on the possibility of synchronizing identical clocks, which assumes a homogeneous and isotropic behaviour of the information carrier, i.e., of light, investigated so far.

This homogeneous and isotropic behaviour of light is possible only in absence of any perturbing factor.

(5) The procedure employed in defining the straight line and distance may be extended to defining the plane, the length of a curve and the angle, and thus the whole of geometry rests on the definitions given above.

Thus, consider two straight lines OP_0 and OM_0 ($O \in S$, $P_0 \in S$; $M_0 \in S$, with O the origin of S) and a point O_1 on OP_0.

Let O_1 be the origin of the space S and D the (infinite and continuous) set of all straight lines Δ determined by O_1 and the current point on line OM_0.

The set of all points $P \in D$ forms the plane determined by points O, P_0 and M_0.

It is obvious that only in an inertial space (as S is) in which the straight line can be defined, may the plane also be defined.

Once distance is defined, one can get to defining the sphere and its intersection with a plane that yields the circle. If the plane Q includes the centre O of sphere σ (O may eventually be the origin of the inertial space S wherein the geometric elements in question are defined), let $C_0 = \sigma \cap Q$ be the intersection circle, and consider two points $A \in C_0$, $B \in C_0$.

The straight lines OA and OB define the <u>angle</u>.

Consider a set of points $A_i \in C_0$ ($i = 1, 2, \ldots, n$) with $A_0 \equiv A$ and $A_n \equiv B$, and let $A_0, A_1, A_2, \ldots, A_{n-1}$ be successively the origins of the inertial space S;

then the distances $l_{i(i-1)} = A_iA_{i-1}$ (with i = 1, 2, ..., n) are established; if ν is the norm of this division, for ν tending to zero, then $l = \sum_{i=1}^{n} l_{i(i-1)}$ is the length of arc AC and $\frac{1}{R} = \phi$ (where R is the sphere radius) is the angle between OA and OB (in radians).

Once these elementary notions are defined in terms of the temporal factor associated with the transmission of information, the whole geometry, the whole structure of space, enjoy a new interpretation, although the usual geometrical truths (axioms, theorems, relations) continue to be valid.

C.I.1.3. Non-inertial spaces. A space whose behaviour is not homogeneous and isotropic (due to influences undergone by the information carrier), and therefore does not allow synchronization of a set of identical clocks (that are assumed to advance at equal rates), is referred to as a non-inertial space.

In such spaces the definitions given above in inertial spaces for straight line and plane do not hold any longer. This may be considered as the physical meaning of curvature of space, a phenomenon studied in the theory of general relativity.

To make this clear, let S be a space which is initially inertial and Δ a straight line OP_0 in this space (O the origin of the space S and $P_0 \in S$) defined as in § C.I.1.2. Assume now that a certain factor interferes, e.g., a gravitational field, which influences (alters) the motion of the information carrier (light) without synchronously altering the behaviour of clocks according to that of light. In the inertial space S, an observer located at $P_1 \in \Delta \subset S$ notes an event E at the same time t regardless of which clock is used: that at P_1, that at O, or that at any point $P_2 \in \Delta \subset S$ (provided that P_2 lies between O and P_1). Due to the perturbation undergone by the information carrier, the observer at P_1 notes the event E' at times t_1, t_0 or t_2 if clocks at P_1, O or P_2, respectively, are watched. Thus, once the perturbation mentioned above exists, the space has ceased to be inertial.

Therefore, according to the definitions in § C.I.1.2, as long as space S was inertial, the points O, P_2, P_1 were colinear; however the colinearity was lost as soon as the perturbing factor (e.g., a gravitational field) interfered and perturbed the information carrier. The Δ geometric variety ($O \in \Delta$; $P_2 \in \Delta$; $P_1 \in \Delta$), a straight line in the inertial space S, ceases to be a straight line in the space which is no longer inertial, following the perturbation of the information carrier.

This is essentially one (but not the only) aspect of the curving of space.

Note. The inertial space is acceptable as an idealization, an approximation. The influence and general interaction that involves the whole universe, the information carriers included, rule out the possibility that there really exists a form of motion, or evaluation (light as well) which does not undergo some form of influence.

Whenever this influence is sufficiently weak, it may be disregarded by idealization or by natural modelling.

However the implications of the space curving are most often so weak that they are very hard to detect.

C.I.1.4. The rigid line. Let C be a body and D a

domain it occupies in space. One restricts the discussion to the case of a domain D assimilated by idealization to a curve Γ and one assumes that:

(a) The ensemble of particles of substance $M \in \Gamma$ that are considered as geometrical points obeys the condition required at § C.I.1.2. in order to lie on a straight line (a line segment).

(b) This property is not lost wherever the body C is placed in space (i.e., $D \equiv \Gamma$) and whatever the influences and interactions it is subjected to.

Then Γ is referred to as a rigid line or, more exactly, a rigid line segment.

It is worth noting that, once use is made of the definition of straight line in § C.I.1.2., the rigid straight line (rigid line segment) may be defined only in an inertial space; if C lies in a non-inertial space, its points cease to obey the conditions of § C.I.1.2., which are compatible with inertial spaces only.

C.I.2. Reference frames

The definitions and considerations above allow a rigorous definition of reference frames and their analysis.

Obviously, it is not important if Cartesian, cylindrical or polar coordinates are employed. For the sake of simplicity, the reasoning is developed in Cartesian systems, extension to cylindrical or polar coordinates being possible.

With the distance between two points and the angle between two lines defined as above, it follows immediately that the distance between a point P and a geometric variety V is the minimum value of the distance between

P and M $\in$ V and may be defined as a Cartesian frame in an inertial space S.

Indeed, consider the origin O of the space S and three non-coplanar straight lines, that in a particular (and the most usual) case intersect in O at right angles with each other. These three straight lines Ox, Oy, Oz, considered according to § C.I.1.2., and oriented in the known manner yield a Cartesian frame. No difficulty is encountered in defining (as outlined above) the x, y, z coordinates of a point P $\in$ S, since they are the distances to planes yOz, xOz, xOy, respectively, and the notions of plane and distance to a geometrical variety (a plane in particular) have already been defined.

C.I.2.1. Inertial frames. Consider a space S and a reference frame Oxyz with O the origin of S_0 space. If the coordinates x, y, z (in the Oxyz frame) of all points P $\in$ S are time-independent (i.e., for any given P $\in$ S, the coordinates x, y, z are constant), then the inertial space S is the space of the Oxyz frame, and this latter is an inertial frame.

The definitions given so far suggest that in this Oxyz frame the laws and relations of Euclidean geometry hold, be they expressed elementarily or analytically.

Thus the coordinates x, y, z of the points of a straight line $\Delta \in$ S in the Oxyz frame satisfy the well-known equations

$$\frac{x - x_1}{x_2 - x_1} = \frac{y - y_1}{y_2 - y_1} = \frac{z - z_1}{z_2 - z_1}, \tag{C.2}$$

where P_{01} (x_1, y_1, z_1) and P_{02} (x_2, y_2, z_2) are two points that determine the line.

Since $\Delta \subset S$, $P_{01} \in \Delta$, and $P_{02} \in \Delta$ it turns out that both P_{01}, P_{02} and any point P_0 are at rest with respect to Oxyz (as is, in fact, the whole space S to which they belong).

Assume the segment $P_{01}P_{02} \subset \Delta$ to be a rigid segment on the straight line and $M_1 \equiv P_{01}$, $M_2 \equiv P_{02}$ (M_1 and M_2 being material points that belong to the rigid segment). For any observer in space S it appears as such.

Assume further that, starting at a certain moment, the M_1M_2 segment moves with respect to the Oxyz frame. Afterwards, all points $M \in M_1M_2$ (that are material points, since they belong to a rigid straight line) cease to belong to the space S.

This statement seems evident if one considers the way that the space S and the frame Oxyz were defined.

The motion of the segment M_1M_2 means that the material points M that make up the segment (M_1 and M_2 are the ends of the rigid line segment formed of points M) occupy successively a continuous set of positions.

Let A be the current position of the rigid segment M_1M_2 in its motion with respect to Oxyz, and let $P \in S$ be the points of the space S which in this position superpose on $M \in M_1M_2$. In other words, in the position in question, $M \equiv P$ (from a geometric viewpoint).

Obviously in another position A' of the rigid segment M_1M_2 ($A' \neq A$), other points $P \in S$ coincide (geometrically) with the material points M.

Since Euclidean geometry holds in the inertial space S, in position A of segment M_1M_2 all points $P \in S$ and $P \neq M$ satisfy the equation:

$$\bar{r} = \bar{r}_1 + \lambda \overline{P_1P_2}, \tag{C.3}$$

where $\bar{r}$ is the position vector of the point $P \equiv M \in M_1M_2$ (and $P \in S$) with respect to O, $\bar{r}_1$ the position vector of $P_1 \equiv M_1$ ($P_1 \in S$), and $\lambda \in [0,1]$ a scalar parameter whose set of values is in a 1:1 correspondence with the set of points $P \equiv M$ ($P \in S$) of the rigid segment M_1M_2. At the same time, $\overline{P_1P_2}$ is the vector defined by P_1 and P_2 ($P_1 \equiv M_1$; $P_2 \equiv M_2$; $P_1 \in S$; $P_2 \in S$).

Eqn. (C.3) is possible just because M_1M_2 is a rigid segment of a straight line (according to the definition given for the rigid segment).

If the information carrier were propagated instantaneously throughout the space S, any observer would have noted the position A of the rigid segment such that points $M \in M_1M_2$ coincided with the set of points $P \in S$, whose positions satisfy the vectorial equation (C.3).

However, the information propagates at finite velocity, i.e., as shown in the chapter dedicated to time, any variation in the universe divided by the space travelled by the information carrier yields a ratio which is not zero.

Consider now an observer in the space S at the point P_1 who watches the rigid segment when it lies in position A ($M_1 \equiv P_1$). If information were propagated instantaneously in the space S, the observer at P_1 would note that the rigid segment in position A occupies a place whose points satisfy the equation:

$$\bar{\rho} = \lambda\overline{P_1P_2}, \qquad (C.4)$$

where $\bar{\rho}$ stands for the position vector of point $P \equiv M \in M_1M_2$ (but $P \in S$ is a geometric point, while M is a material one ($M \notin S$) in motion with respect to S).

In actuality, propagation of light is not instantaneous, and the observer at P_1 notes the points M as lying on a geometric variety whose equation is:

$$\bar{\rho}' = \bar{\rho} - \bar{w} = \lambda\overline{P_1P_2} - \bar{w}, \tag{C.5}$$

where $\bar{w}$ is the distance covered by M, while information travels from P' to P_1. Obviously $\overline{P'P} = \bar{w}$. In other words the observer at P_1 sees the set of points $M \in M_1M_2$ as coinciding with the set of geometrical points $P' \in S$ whose position vectors satisfy eqn. (C.5), rather than with the set of geometrical points $P \in S$ that satisfy eqn. (C.4) owing to non-instantaneous transport of information.

Therefore one can ask a natural question: what condition should be fulfilled in order that eqn. (C.5) stands for a line segment?

Evidently this line segment has to include the point $P_1 \equiv M_1$ that lies at its terminal point. The equation would be:

$$\bar{\rho}' = \lambda\overline{P_1P_2'}, \tag{C.6}$$

where P_2' is the point from which observer P_1 sees particle M_2 at the other terminal point of the rigid segment M_1M_2.

However eqn. (C.6) is a particular case of eqn. (C.5), i.e., the case where $\bar{w} = w\,\bar{u}_0$, $\bar{u}_0$ being a constant unit vector and

$$\bar{\rho}' = \lambda\overline{P_1P_2} - \bar{w} = \lambda\overline{P_1P_2} - w\,\bar{u}_0 =$$

$$\lambda(\overline{P_1P_2} - q_0u_0), \tag{C.7}$$

where q_0 stands for a scalar constant. It follows that $\overline{P_1P_2} - q_0\bar{u}_0 = \overline{P_1P_2'}$.

It turns out that the observer at $P_1 \in S$ sees the rigid segment M_1M_2 take up a position that coincides with that of a line segment in the space S if $\bar{w} = \lambda q_0\bar{u}_0$ (although the observation may be influenced by the non-instantaneous transport of information).

On the other hand $\mu = \lambda|\overline{P_1P_2}|$ represents the distance covered by information (Fig. C-1) from P' to P_1. (To each λ value corresponds a material point $M \in M_1M_2$ and a point $P' \in S$; however $P' \in M$, according to the findings of the observer at P_1.)

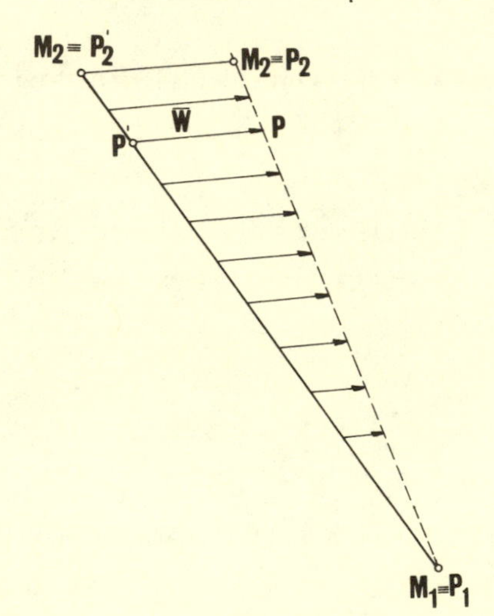

Fig. C-1.

Thus:

$$\frac{|\bar{w}|}{\mu} = \frac{\lambda q_0}{\lambda|\overline{P_1P_2}|} = \frac{q_0}{|\overline{P_1P_2}|} = \text{constant.} \qquad (C.8)$$

This <u>ratio</u> is constant because $|\overline{P_1P_2}|$ represents the length of the rigid segment measured at rest (before its motion starts) or, what is the same, the proper length of the segment, and the constant q_0 depends on the existence of eqn. (C.6).

It is easy to see that eqn. (C.8) signifies that the value of the velocity of any point M is the same and constant.

Indeed the ratio of any variation over the distance covered by light represents the velocity of that variation expressed directly in terms of the timing phenomenon supplied by the propagation of information.

The direction and sense of velocities of the points M are expressed by the unit vector $\bar{u}_0$; this latter is also constant (condition for eqn. (C.6) to exist), and this shows that the direction of velocities of the points M is constant, as well.

Conclusion. An observer who looks along a rigid line segment sees it as a line segment even if in motion, provided that:

(a) the observer resides in an inertial space S;

(b) motion of the segment is considered with respect to such an inertial space S;

(c) motion of the rigid segment belongs to the same equivalence class as the transmission of information in an inertial space S. This motion is referred to as rectilinear and uniform translation. The velocities of all material particles of the segment are constant vectors with respect to Oxyz (i.e., with respect to the space S).

If at least one of the conditions stated is not fulfilled, the observer sees the rigid segment to have lost the rectilinear character. It is then said that the segment has curved.

It follows that: (1) If the conditions stated are not fulfilled, the notion of a rigid segment loses sense; (2) the physical sense of one of the aspects of space curvature is evident.

In fact, a similar phenomenon was encountered in the presentation of relativistic mechanics. Nevertheless there is a difference at least of emphasis, between the shape change of a body, on passing from one frame to the other by means of Lorentz transformations, and the curving of a line segment whenever its motion with respect to inertial space S does not fulfill the conditions stated, although the inner cause of both is the transmission of information.

Indeed, if the segment M_1M_2 is engaged in a motion (with respect to Oxyz, i.e., S) that is not uniform and rectilinear translation, even an observer who is attached to the segment (e.g., is always bound to M_1) will see it as a curved segment.

The rectilinear uniform translation belongs, as already mentioned, to the same equivalence class as the propagation of information. Because the propagation of information was selected to be the timing motion (see the chapter dealing with time), it follows that in rectilinear uniform translation the velocities of all points are constant. Obviously these velocities refer to the motion with respect to Oxyz or, what is the same, to the inertial space S.

The constant velocity motion with respect to an inertial space is referred to as an inertial motion.

Consider a space σ whose points move at constant velocity, the same for any point $Q \in \sigma$ and invariable in time with respect to inertial space S.

According to the foregoing, any rigid line segment

S remains a line segment in σ as well. The velocity of propagation of information which is constant in space S is also constant in space σ, because stating that all points of σ move with constant velocity with respect to S amounts to saying that the motion of σ with respect to S belongs to the same equivalence class as the motion of the information carrier (light) with respect to S.

Since the two motions (of space σ and of light) belong to the same equivalence class, the speed of light appears to be constant in σ, as well, and the information transmission may be (even should be) chosen as timing phenomenon in space σ as well as in space S; all motions in these spaces are referred to the information transmission whose speed (referred to itself) both in S and σ appears to be evidently equal to unity. Or else, a change of scale which is equivalent to adopting an abstract t (time) variable instead of the real one representing the space travelled along by information (the distance covered by light), yields a speed of propagation given by the same constant both in S and in σ.

Even this brief analysis allows one to realize that spaces S and σ are in perfectly identical situations. Consequently, not only S, but the space σ is also an inertial space wherein propagation of information is homogeneous and isotropic.

It follows that, in order to be allowed to define a straight line in σ, one has to choose a Cartesian frame $O_1x_1y_1z_1$ bound to σ; O_1 is the origin of the space σ (the synchronization of clock in σ is achieved with respect to O_1) and O_1x_1, O_1y_1, O_1z_1 are the orthogonal axes. Now, synchronization of clocks in σ is possible since the propagation of information (light) is homogeneous and isotropic (in σ as in S).

σ is the space of the $O_1x_1y_1z_1$ frame, because the coordinates of any point $Q \in \sigma$ are constant in $O_1x_1y_1z_1$, and therefore it is at rest with respect to $O_1x_1y_1z_1$.

The frame $O_1x_1y_1z_1$ is an inertial frame, as is Oxyz.

However, it is obvious that the two inertial frames Oxyz and $O_1x_1y_1z_1$ are not at rest with respect to motion in such a way that all their points move with constant velocity ($P \in S$ with respect to $O_1x_1y_1z_1$ and $Q \in \sigma$ with respect to Oxyz).

The relative motion of frames Oxyz and $O_1x_1y_1z_1$ belongs to the same equivalence class with propagation of information in inertial spaces. It is an inertial motion.

In conclusion. Any frame with respect to which propagation of information is homogeneous and isotropic (that is, is not subject to any perturbing action) is an inertial frame. The inertial frames move with respect to each other (in particular they may be at rest), and their relative motion, called inertial motion, belongs to the same equivalence class with the propagation of information (light) when this latter is not perturbed.

Such a motion is called rectilinear uniform motion.

The transformation of space-time coordinates in transfer from one inertial frame to the other (also inertial) is made according to the Lorentz-Einstein group.

Note. Considering the interaction, the general influence, the existence in the universe of all forms of motion (the motion of information carrier, i.e., of the wave or form of energy that carries information), one has to conclude that this concept of inertial space and, implicitly, of inertial frame results from an idealization, an approximation the closer to truth, the lower the influences that perturb the propagation of information in the space assumed to be inertial.

C.I.2.2. Non-inertial frames. Any frame with respect to which the transmission of information is non-homogeneous (differs from point to point) or non-isotropic is called a non-inertial frame.

The non-inertial character of the frame may be determined by the non-inertial nature of the space whose points are referred to the frame in question or by the non-inertial motion of the frame with respect to an inertial space; both these factors may coexist, which causes the frame considered to be non-inertial.

Non-inertial motion is that motion which does not belong to the same equivalence class as the uninfluenced propagation of information. Taking into account that any clock has to belong to the equivalence class of information propagation, i.e., that the speed of light is constant, it follows that the non-inertial motion may not be of constant speed. It is characterized by the presence of accelerations.

Notes. (1) If in very special circumstances, in non-inertial spaces, some particles M undergo such influences that their motion belongs to the same equivalence class as the propagation of information (which is influenced as well, since the space is non-inertial), i.e., the parameters of the M particles' motion and those of information propagation differ by constant factors of proportionality, for these M particles (and only for them) the non-inertial space in question behaves as an inertial space.

(2) In non-inertial spaces or in non-inertial motions,

the motion of straight line and that of a plane may not be defined in the sense adopted in C.I.1.2., and therefore even the notion of a non-inertial frame seems inadequate because its constituent elements may not be rigorously defined. A. Einstein found the solution offered by the utilization of Gaussian coordinates (which bring about the so called Einstein's "mollusc").

Another way of resolving these difficulties would be to consider some sufficiently small domains so that within them the effects of space curving may be disregarded, in other words, to consider a local frame at every point of a non-inertial space.

(3) If the effect of space curving as a result of the non-instantaneous propagation of information is disregarded (or if the propagation of information is considered instantaneous, which is the same), one returns to the classical prerelativistic concept of Newtonian mechanics. In this hypothesis the non-inertial frames may be defined as rigorously as the inertial ones, regardless of whether they involve Cartesian, cylindrical or spherical coordinates. However, the rougher the approximation admitted, the more pronounced the non-inertiality.

(4) Finally, another way of defining <u>non-inertial</u> frames is possible if one resorts to the notion of line segment. Thus, one may accept that for two points $A \in S$ and $B \in S$ (where S is a non-inertial space) there exists a set of all rigid line segments that may take, at a certain moment, such a position that their terminations coincide with A and B; these segments are equal to each other, and it is agreed that they are considered equal even in non-inertial spaces where such segments cease to appear as rectilinear.

Hence, one may define the <u>rigid body</u> both in inertial and non-inertial spaces, in inertial and non-inertial motion.

A rigid body is called that body C_0 for which any pair of particles $M_1 \in C_0$ and $M_2 \in C_0$ may be the ends of a rigid line segment.

It is obvious that:

(a) If the rigid body C_0 is engaged in an uniform rectilinear motion with respect to an inertial space, its shape will seem invariable to any observer in that space. Then there is an inertial frame with respect to which C_0 is at rest. This is the <u>proper frame</u> of the rigid body in question.

(b) The observers located in various inertial frames, in relative (inertial) motion, note various shapes of C_0 (which is also assumed to be in inertial motion) although each observer considers it nondeformable. (E.g., an observer in space S sees a cube, that in space σ, a parallelepiped, but neither the cube nor the parallelepiped seem deformable to the observers.) This relativistic effect is well known.

(c) A rigid body C_0 which lies in a non-inertial space or is engaged in a non-inertial motion is perceived by an observer as a variable-shape body. This statement may seem to conflict with the notion of rigid body.

However, if the definition of rigid body is admitted, it turns out that, although some observers note the shape variation of C_0 under the conditions specified, when the body is returned into an inertial space, the observers with respect to whom the motion of the body C is inertial note that it retains its shape, no matter what interactions it is subjected to.

Thus, the rigid body is defined as that body whose shape does not ever change because of mechanical interactions it may be subjected to.

If an observer notes changes of the geometry of the body, they must be related to the way motion of information propagation (from the body to the observer) is related to the relative motion of the body and the observer.

Obviously various observers participating in various relative (non-inertial) motions note that the body C_0 (rigid in the sense now defined) assumes various shapes. If the observers were able to communicate their findings regarding the geometry of the body, they would not be identical. This means that such a body undergoes an apparent deformation.

Thus, a relativity of the shape is present, but it does not exclude the notion of proper shape: that of the rigid body lying at rest in an inertial frame. Admitting that this proper shape does not change, regardless of the interactions of C_0 with other bodies, one gets to the rigidity of the body. As known, the concept of a rigid body is an idealization. There is no rigid body (which is not deformed by interactions). However in many instances, changes of shape are negligible whatever the interactions.

If a pair of points A_1 and A_2 of a space R may always be connected by a rigid line segment, space R is referred to as a rigid space. If such a rigid space is in inertial motion, any observer in an inertial space notices that the distances between the points of space R are constant. Then R is also an inertial space. If now a (rectangular Cartesian) reference frame Oxyz is attached to R and the points of the axes are maintained constant after R has ceased to be in inertial motion, the coordinates of any $A \in R$ continue to be constant in Oxyz, even in the non-inertial case, provided that distances are measured in R (in Oxyz, implicitly) with rigid line segments.

(5) The reciprocal inertial motion of two frames is readily conceived when both are inertial.

In case of non-inertial frames, their rectilinear and uniform translation with respect to each other is acceptable only within sufficiently small portions, i.e., within infinitesimal domains.

C.II. THE INERTIA

Any mechanical motion, whatever the level of precision of its study, i.e., no matter whether the point of view of classical, Newtonian or relativistic mechanics is adopted, has to be referred to a frame.

This truth, although trivial, raises some subtle problems related to: (a) selection of frame; (b) watching the motion with respect to the selected frame and expressing it quantitatively. This is so because of the transmission of information, which mediates any observation and measurement. If the phenomenon of transmission of information is disregarded and propagation is considered instantaneous, one returns to the classical concept, which is much simpler, but then the results are only approximations to reality.

C.II.1. Aspects related to motion of particles in inertial and non-inertial spaces

Both the problems of selection of frame (which is an

inertial frame) and that of expressing the space-time coordinates of an event (implicitly of expressing mechanical motion) are considerably simpler in inertial spaces than in non-inertial spaces. These problems are solved by general relativity.

The mechanical motion of one particle involves a continuous set of events, each event consisting of the presence of a particle in a certain point of space at the time it occurs.

The mechanical motion of a material system consists of the union of motions of constituent particles.

In inertial frames, because of the possibility of synchronizing clocks, selection of frame and expressing the space-time coordinates by any observer are readily possible. The possibility of synchronizing clocks is offered, as shown above, by the uninfluenced propagation of information.

C.II.1.1. General aspects of the study of motion in non-inertial spaces. The relativistic mechanics which was presented in the first part of this book supplies the method for study of motions with respect to inertial frames.

However, some additional aspects may require further consideration.

Thus, consider $\bar{v}$, the velocity of a particle (with respect to an inertial frame Oxyz) of mass m, at time t (in the same frame) and $\bar{h} = m\bar{v}$ and $E = mc^2$, the momentum and kinetic energy of the particle, respectively; then it follows that:

$$m = \frac{m_0}{\sqrt{1 - \frac{v^2}{c^2}}}, \tag{C.9}$$

where m_0 is the rest mass, c the speed of light and

$$\bar{v} \cdot d\bar{h} = dE, \tag{C.10}$$

may be considered as the fundamental equation of dynamics both for relativistic and classical mechanics.

If the rate of variation of the momentum,

$$\bar{F} = \frac{d\bar{h}}{dt}, \tag{C.11}$$

referred to as force, is known, then:

$$\frac{d}{dt}(m\bar{v}) = \bar{F} \tag{C.12}$$

becomes the differential equation of motion of the particle in Oxyz, no matter whether the study is conducted at the level of precision of relativistic mechanics or within the approximation made according to Newtonian mechanics.

But $\Delta\bar{h} \neq 0$ implies the existence of an interaction.

The principles of mechanics might all be replaced by the following axiom:

Any variation of the momentum of a particle M originates in the exchange of momentum between the particle M and at least one other particle. The rate of momentum variation of a particle is referred to as force, and the exchange of momentum is the mechanical interaction between the particles in question.

Thus, if a particle M does not interact with any particle, $d\bar{h} = 0$ and, according to (C.10), it is required that $dE = 0$, i.e., $E = E_0$ = constant or:

$$\frac{m_0 c^2}{\sqrt{1 - \frac{v^2}{c^2}}} = E_0, \qquad (C.13)$$

and hence that

$$\bar{v} = \frac{c}{E_0} \sqrt{E_0^2 - m_0^2 c^4} = \text{const.} \qquad (C.14)$$

At the same time, $d\bar{h} = 0$ means $m\bar{v} = \bar{h}_0$ = constant, which amounts to:

$$\bar{v} = \frac{\bar{h}_0}{m} = \frac{\bar{h}_0 c^2}{E_0} = \text{const.} \qquad (C.15)$$

It follows that, in the case when a material point is not subject to any interaction, its velocity is a constant vector and, consequently, in an inertial space, it moves rectilinearly and uniformly. This truth is expressed in classical mechanics as its first principle.

In the case when the particle M interacts concomitantly with n other particles, $P_1, P_2, \ldots, P_n$ which brings about momentum variations $d\bar{h}_1, d\bar{h}_2, \ldots, d\bar{h}_n$ in time dt, then $d\bar{h} = \sum_{i=1}^{n} d\bar{h}_i$, and eqn. (C.12) becomes:

$$\frac{d}{dt}(m\bar{v}) = \sum_{i=1}^{n} \frac{d\bar{h}_i}{dt} \qquad (C.16)$$

or else:

$$\frac{dm}{dt}\bar{v} + m\frac{d\bar{v}}{dt} = \sum_{i=1}^{n} \frac{d\bar{h}_i}{dt} = \sum_{i=1}^{n} \bar{F}_i, \qquad (C.17)$$

which is equivalent to <u>the second principle</u>, even identical with it if one neglects the relativistic variation of mass ($\frac{dm}{dt} = 0$).

If two particles M and P interact, then, with equivalent notations, one may write:

$$\bar{v}_P \cdot d\bar{h}_P = dE_P, \ \bar{v}_M \cdot d\bar{h}_M = dE_M. \tag{C.18}$$

Assuming that M interacts only with P and P only with M, the laws of conservation of energy and momentum yield:

$$dE_P + dE_M = 0, \ d\bar{h}_P + d\bar{h}_M = 0. \tag{C.19}$$

The equalities remain true in classical mechanics on the basis of the equation:

$$\frac{d\bar{h}_P}{dt} = \frac{d\bar{h}_M}{dt}, \tag{C.19'}$$

and together with:

$$\frac{d\bar{h}_P}{dt} = \bar{F}_{MP}, \ \frac{d\bar{h}_M}{dt} = \bar{F}_{PM}, \tag{C.20}$$

yield $\bar{F}_{MP} = -\bar{F}_{PM}$, that is, the <u>third principle</u>, which is the principle of the equality between active force (e.g., that exerted by M upon P, i.e., $\bar{F}_{MP}$) and the reaction ($\bar{F}_{PM}$ exerted by P upon M).

As the two particles generally do not touch and the mutual influence propagates non-instantaneously in space, eqns. (C.19) and (C.19') are not exact.

It follows that, if the long range interaction propagates at finite speed, the principle of equality between action and reaction holds only approximately except for the case of direct contact (zero distance) between interacting particles.

C.II.1.2. Mescerski-Levi-Civita equation in the relativistic case. The case of the material point whose mass changes not only relativistically but also by capture or emission of some (continuous or discontinuous) jets of particles has attracted outstanding interest, because of practical applications and because it allows the possibility of a rigorous treatment (a direct contact is involved and therefore the time required for propagation of mutual information between interacting particles is zero).

Emission is treated as negative capture and capture as a plastic collision between the material point and the particles captured.

Consider m_0, the rest mass of the material point M which captures a jet of particles. Obviously this mass is time-dependent ($m_0 = m_0 (t)$). Let $v = v(t)$ be the velocity of point M with respect to the inertial frame motion is referred to and $u = u(t)$ denote the velocity of captured particles at the time of capture (more precisely at a time just before capture).

The phenomenon is studied throughout the interval $[t, t + \Delta t]$ whose duration Δt is assumed to be sufficiently small for the following equations (which are a vectorial form of eqns. (A.142) and (A.143) combined with (C.10) and (C.12)) to hold:

$$\bar{F}\Delta t + \frac{m_0 \bar{v}}{\sqrt{1 - \frac{v^2}{c^2}}} + \frac{\Delta m \bar{u}}{\sqrt{1 - \frac{u^2}{c^2}}} =$$

$$\frac{(m_0 + \Delta m_0)(\bar{v} + \Delta\bar{v})}{\sqrt{1 - \frac{(\bar{v} + \Delta\bar{v})^2}{c^2}}}, \qquad (C.21)$$

$$\bar{F}.\bar{v}\,\Delta t + \frac{m_0 c^2}{\sqrt{1 - \frac{v^2}{c^2}}} + \frac{\Delta m c^2}{\sqrt{1 - \frac{u^2}{c^2}}} =$$

$$\frac{(m_0 + \Delta m_0) c^2}{\sqrt{1 - \frac{(\bar{v} + \Delta\bar{v})^2}{c^2}}}. \qquad (C.22)$$

In these equations Δm is the rest mass of captured (emitted) particles throughout the interval $[t, t + \Delta t]$; F is the force acting upon M completely independent of capture (emission). That is, it measures certain interactions; $\Delta\bar{v}$ is the variation of $\bar{v}$ (of point M) in the time interval Δt; Δm_0 is the variation of rest mass m_0.

Eqn. (C.21) may be rewritten as:

$$m_0 \left[\frac{\bar{v} + \Delta\bar{v}}{\sqrt{1 - \frac{(\bar{v} + \Delta\bar{v})^2}{c^2}}} - \frac{\bar{v}}{\sqrt{1 - \frac{v^2}{c^2}}} \right] +$$

$$\frac{\Delta m_0 \; (\bar{v} + \Delta\bar{v})}{\sqrt{1 - \frac{(\bar{v} + \Delta\bar{v})^2}{c^2}}} = \bar{F}\Delta t + \frac{\Delta m \bar{u}}{\sqrt{1 - \frac{u^2}{c^2}}} \qquad (C.23)$$

or

$$\frac{m_0\Delta\bar{v} + \Delta m_0\bar{v}}{\sqrt{1 - \frac{(\bar{v} + \Delta\bar{v})^2}{c^2}}} +$$

$$m_0\bar{v}\left[\frac{1}{\sqrt{1 - \frac{(\bar{v} + \Delta\bar{v})^2}{c^2}}} - \frac{1}{\sqrt{1 - \frac{v^2}{c^2}}}\right] +$$

$$\frac{\Delta m_0\Delta\bar{v}}{\sqrt{1 - \frac{(\bar{v} + \Delta\bar{v})^2}{c^2}}} = \bar{F}\Delta t + \frac{\Delta m\bar{u}}{\sqrt{1 - \frac{u^2}{c^2}}}. \tag{C.23'}$$

Considering now the limiting case when Δt tends to zero ($\Delta t \to 0$) and implicitly Δm, Δm_0, and $\Delta\bar{v}$ also tend to zero ($\Delta m \to 0$, $\Delta m_0 \to 0$, $\Delta\bar{v} \to 0$), one notices that:

$$\lim_{\Delta t \to 0} (m_0\Delta\bar{v} + \Delta m_0\bar{v}) = d\,(m_0\bar{v}), \tag{C.24}$$

$$\lim_{\Delta t \to 0}\left[\frac{1}{\sqrt{1 - \frac{(\bar{v} + \Delta\bar{v})^2}{c^2}}} - \frac{1}{\sqrt{1 - \frac{v^2}{c^2}}}\right] =$$

$$d\left[\left(1 - \frac{v^2}{c^2}\right)^{-\frac{1}{2}}\right] = \frac{v\,dv}{c^2\sqrt{\left(1 - \frac{v^2}{c^2}\right)^3}} =$$

$$\frac{d\left(\frac{v^2}{c^2}\right)}{2\sqrt{\left(1 - \frac{v^2}{c^2}\right)^3}}, \tag{C.25}$$

$$\lim_{\Delta t \to 0} \frac{1}{\sqrt{1 - \frac{(\bar{v} + \Delta\bar{v})^2}{c^2}}} = \frac{1}{\sqrt{1 - \frac{v^2}{c^2}}} +$$

$$\frac{vdv}{c^2\sqrt{\left(1 - \frac{v^2}{c^2}\right)^3}} = \frac{1}{\sqrt{1 - \frac{v^2}{c^2}}} +$$

$$\frac{d\left(\frac{v^2}{c^2}\right)}{2\sqrt{\left(1 - \frac{v^2}{c^2}\right)^3}}, \tag{C.25'}$$

$$\lim_{\Delta t \to 0} \Delta m_0 \Delta\bar{v} = 0. \tag{C.26}$$

And combining eqns. (C.24), (C.25), (C.25') and (C.26) with eqn. (C.23') (this latter for $\Delta t \to 0$) one gets:

$$\frac{d(m_0\bar{v})}{\sqrt{1 - \frac{v^2}{c^2}}} + \frac{d(m_0\bar{v})\ d\left(\frac{v^2}{c^2}\right)}{2\sqrt{\left(1 - \frac{v^2}{c^2}\right)^3}} + \frac{m_0\bar{v}\ d\left(\frac{v^2}{c^2}\right)}{2\sqrt{\left(1 - \frac{v^2}{c^2}\right)^3}} =$$

$$\bar{F}dt + \frac{\bar{u}dm}{\sqrt{1 - \frac{u^2}{c^2}}}, \tag{C.27}$$

with the already known meaning of symbols.

Neglecting the higher-order infinitesimals, i.e., admitting that $d(m_0\bar{v}) \cdot d(\frac{v^2}{c^2}) \cong 0$, and considering eqn. (C.25), again yield:

$$\frac{d}{dt}\left(\frac{m_o\bar{v}}{\sqrt{1-\frac{v^2}{c^2}}}\right) = \bar{F} + \frac{\bar{u}}{\sqrt{1-\frac{u^2}{c^2}}}\frac{dm}{dt}. \qquad (C.28)$$

This is the basic equation of the dynamics of a material point with variable mass, i.e., the Mescerski-Levi-Civita equation for the relativistic case; in this case, to the variation of mass by capture (emission) of a jet of particles is added the relativistic variation of the mass that captures (emits) and of the captured (emitted) masses.

It is noteworthy that:

(1) For $v^2c^{-2} \cong 0$ one returns to the well-known classical form of the equation, which is natural.

(2) In case of capture, $\Delta m > 0$ and thus $\frac{dm}{dt} > 0$, while in case of emission, $\Delta m < 0$, which means $\frac{dm}{dt} < 0$.

(3) Both m_0 and dm are <u>rest masses</u> (m_0 captures or emits while dm is captured or emitted) and if, in time dt, m_0 varies with the elementary amount dm_0, in general $dm_0 \neq dm$, which is evident from (C.28).

Insertion of eqns. (C.24), (C.25), (C.25'), (C.26) in (C.22), the latter being also particularized for Δt tending to zero, yields:

$$\bar{F}\cdot\bar{v} + \frac{c^2}{\sqrt{1-\frac{u^2}{c^2}}}\frac{dm}{dt} = c^2\frac{d}{dt}\left(\frac{m_0}{\sqrt{1-\frac{v^2}{c^2}}}\right). \qquad (C.29)$$

This energy equation gives immediately:

$$\frac{d}{dt}\left(\frac{m_0}{\sqrt{1-\frac{v^2}{c^2}}}\right) = \frac{\bar{F}\cdot\bar{v}}{c^2} + \frac{\frac{dm}{dt}}{\sqrt{1-\frac{u^2}{c^2}}}, \qquad (C.29')$$

and because:

$$m_M = \frac{m_0}{\sqrt{1 - \frac{v^2}{c^2}}}, \tag{C.30}$$

eqn. (C.29') may be rewritten:

$$dm_M = \frac{\bar{F}\cdot\bar{v}}{c^2}\,dt + \frac{dm}{\sqrt{1 - \frac{u^2}{c^2}}} \tag{C.31}$$

or:

$$\frac{dm_0}{\sqrt{1 - \frac{v^2}{c^2}}} + m_0 d\left(\frac{1}{\sqrt{1 - \frac{v^2}{c^2}}}\right) =$$

$$\frac{\bar{F}\cdot\bar{v}}{c^2}\,dt + \frac{dm}{\sqrt{1 - \frac{u^2}{c^2}}}. \tag{C.31'}$$

On the other hand, if $\bar{F}\cdot\bar{v} \cong 0$ (a particular yet interesting case), one can write:

$$d\left(\frac{m_0}{\sqrt{1 - \frac{v^2}{c^2}}}\right) = \frac{dm}{\sqrt{1 - \frac{u^2}{c^2}}}. \tag{C.32}$$

Both eqns. (C.31') and (C.32) give $dm_0 = dm$ only if $v^2c^{-2} \cong 0$ and $u^2c^{-2} \cong 0$. If at least one of the inequalities $v^2/c^2 > 0$, $u^2/c^2 > 0$ is satisfied, $dm_0 \neq dm$. The physical interpretation of this fact is that the product of rest mass m_0 with the square of the speed of light represents the internal energy of the particle; the rest mass varies on capture (emission) of particles not only by addition (elimination) of captured (emitted) mass but by the variation of internal energy following capture (emission).

C.II.1.3. The effect of percussion force in the relativistic case. Consider a percussion force:

$$\bar{F} = \bar{P}\,\delta(t - \tau), \tag{C.33}$$

where $\bar{P}$ is the percussion vector, τ the time of percussion and $\delta\,(t - \tau)$ the Dirac distribution. The collision is naturally assumed to be ideal (instantaneous).

Application of this force to a mass point M, whose rest mass is m_0 (according to (C.12)), yields:

$$\frac{d}{dt}\,(m\bar{v}) = \bar{P}\,\delta(t - \tau), \tag{C.34}$$

a differential equation, which expresses motion of M of velocity $\bar{v}$ with respect to an inertial frame.

Integration (in the sense of distribution theory) yields:

$$m\bar{v} = \bar{P}\,H(t - \tau) + \bar{K}_0, \tag{C.35}$$

where:

$$H(t - \tau) = \begin{cases} 0 \text{ for } t < \tau \\ \tfrac{1}{2} \text{ for } t = \tau \\ 1 \text{ for } t > \tau \end{cases} \tag{C.36}$$

is the Heaviside step function; $\bar{K}_0$ is an arbitrary constant whose significance is evident.

Indeed one can see that:

$$\lim_{t \to \tau \to 0} (m\bar{v}) = K_0 = m'\bar{v}', \qquad (C.37)$$

$\bar{v}'$ and m' being the velocity and mass of particle M before application of the percussion force $\bar{F}_p$ (i.e., for $t < \tau$ or more exactly $t = \tau - 0$).

Thus (C.35) means that:

$$m\bar{v} = \bar{P}\, H(t - \tau) + m'\bar{v}', \qquad (C.38)$$

or, in more detail:

$$\frac{m_{0p}\bar{v}}{\sqrt{1 - \frac{v^2}{c^2}}} = \bar{P}\, H(t - \tau) + \frac{m_0'\bar{v}'}{\sqrt{1 - \frac{v'^2}{c^2}}}, \qquad (C.39)$$

with m_{0p} the rest mass after collision (because the rest energy may rise, $m_{0p} > m_0'$; only for elastic collision is $m_{0p} = m_0'$); obviously for $t > \tau$, $m_0 = m_{0p}$ and, for $t < \tau, m_0 = m_0'$.

Immediately after collision the particle M moves with a velocity $\bar{v}_1$ which satisfies the equation:

$$\frac{m_{0p}\bar{v}_1}{\sqrt{1 - \frac{v_1^2}{c^2}}} = \bar{P} + \frac{m_0'\bar{v}'}{\sqrt{1 - \frac{v'^2}{c^2}}} \qquad (C.40)$$

as a consequence of eqn. (C.36).

Obviously m_0 is the rest mass before collision.

The problem arises to establish the effect of a percussion in the presence of the relativistic variation of mass with velocity.

Because of percussion $\bar{P}$, the velocity jumps with $\bar{v} = \bar{v}_1 - \bar{v}'$. One adopts the notation:

$$\bar{Q} = \frac{\bar{P}}{m_0'} + \frac{\bar{v}'}{\sqrt{1 - \frac{v'^2}{c^2}}}, \tag{C.41}$$

where $\bar{Q}$ is known and depends exclusively on the data of the problem.

Substitution of (C.41) into (C.40) and routine calculations yield:

$$\bar{v}_1 = \frac{\frac{m_0'}{m_{0p}} \bar{Q}}{\sqrt{1 + \frac{m_0'^2 Q^2}{m_{0p}^2 c^2}}} \tag{C.42}$$

and consequently the velocity jump:

$$\Delta\bar{v} = \bar{v}_1 - \bar{v}' = \frac{\frac{m_0'}{m_{0p}} \bar{Q}}{\sqrt{1 + \frac{m_0'^2 Q^2}{m_{0p}^2 c^2}}} - \bar{v}' =$$

$$\frac{\bar{P}}{m_{0p} \sqrt{1 + \frac{m_0'^2 Q^2}{m_{0p}^2 c^2}}} +$$

$$\bar{v}'\left(\frac{m_0'}{m_{0p}\sqrt{1-\frac{v'^2}{c^2}}\sqrt{1+\frac{m_0'^2 Q^2}{m_{0p}^2 c^2}}}-1\right), \quad \text{(C.43)}$$

is thus determined for the (relativistic) case considered.

It is worth noting that:

(1) Vector $\bar{Q}$ has the dimension of a velocity;

(2) If $v'^2 c^{-2} \cong 0$, $\bar{Q} \cong m_0'^{-1} \bar{P} + \bar{v}$;

(3) Moreover, if $Q^2 c^{-2} \cong 0$ and $m_0' \cong m_{0p}$, then $\bar{v} = \frac{P}{m_0'}$, which is the well-known equality in classical mechanics, which expresses the velocity jump determined by a percussion $\bar{P}$;

(4) the m_{0p} mass satisfies the equality $m_{0p}c^2 = m_0'c^2 + \Delta E_0$, ΔE_0 being the increase of internal energy of the particle following collision. When the collision is elastic $\Delta E_0 = 0$ and $m_{0p} = m_0'$.

If a succession of percussions is applied to a particle at times τ_i, during the action of a continuous force $\bar{F}$, the differential law of motion is:

$$\frac{d}{dt}\left[\frac{m_0\bar{v}}{\sqrt{1-\frac{v^2}{c^2}}}\right] = \bar{F} + \sum_{i=1}^{n} \bar{P}_i \delta(t-\tau_i), \quad \text{(C.44)}$$

where n is the total number of percussions.

The velocity jumps caused by percussions are calculated step by step by eqn. (C.43).

It may easily be seen that eqn. (C.43) is a generalization of the classical one which makes it suitable for the case with relativistic variation of mass.

C.II.2. Manifestation of inertia

Inertia is briefly analyzed next; a new presentation and mostly a new interpretation of the phenomenon are attempted.

C.II.2.1. <u>The inertial spaces</u>. Consider a particle that moves in an inertial space without suffering any influence; this motion referred to an inertial frame is considered a timing motion and therefore belongs to the same equivalence class as the propagation of information (light). In order that motion of a particle with respect to an inertial frame is not equivalent to that of information (light), the particle has to interact with other particles, i.e., has to be imparted a momentum <u>different from zero</u>.

This is a manifestation of the property that is referred to as <u>inertia</u> whenever motion occurs in an inertial space (previously defined).

Eqn. (C.10), through its general and fundamental character, expresses in a concise quantitative form the inertia of the particle it refers to.

Indeed, assume a particle M, whose rest mass is m_0, interacts with n other particles that transfer to it in time dt the momentum $d\bar{h}_s$ (s = 1, ..., n) each; the momentum variation of the mass point M is $d\bar{h} = \sum_{s=1}^{n} d\bar{h}_s$. If M is a single particle (does not interact with any other material system), i.e., if n = 0 then $d\bar{h} = 0$. It is possible that although $d\bar{h}_s \neq 0$ (s = 1, 2, ..., n), $d\bar{h} = \sum_{s=1}^{n} d\bar{h}_s = 0$, the particle M behaving as an isolated one. It was shown that in such a case (eqns. (C.13), (C.15)) the velocity is constant, which amounts to saying that M belongs to the equivalence class of the propagation of information.

C.II.2.2. Non-inertial spaces. In a non-inertial space the factors that influence the motion of the information carrier (light) influence any other motion of any material system.

Let σ be such a non-inertial space.

As shown above, the synchronization of clocks and implicit defining of a straight line are not possible in such a space.

There is however a particular situation when, in spite of the non-inertiality of space, synchronization of clocks is possible. This is the case of rigid spaces (in the sense defined in C.I.2.2.). However, for synchronization of clocks in such a space, one has to use clocks whose period differ from point to point; the timing parameter of the clock at a point $M \in \sigma$ belongs to the same equivalence class as the propagation of information (light) at the point M. (This propagation is not necessarily uniform throughout the whole space σ which is non-inertial.)

In other words, synchronization of clocks in a non-inertial space σ is possible if (and only if) at any point $M \in \sigma$ one may place a clock whose parameter varies equivalently with propagation of information at that point M. However the non-inertiality of σ may require that, in these circumstances, clocks at various points $M_i \in \sigma$ $(i = 1, 2, \ldots, n)$ are such that, when all brought together at M_0, they do not show equal variations of their parameters between two events E_1 and E_2, i.e., they do not have all the same rhythm. Nevertheless, because there is a perfect equivalence between the parameter of the clock at M_i and the propagation of information, no matter what the points $M \in \sigma$, it turns out that light permanently undergoes an influence at any point $M \in \sigma$, an influence that may differ from one point to the other (the influence at M_j may differ from that at M_k, $M_j \neq M_k$, if $j \neq k$).

It is possible that, if one chooses adequate rhythms and suitably delays the origin of clocks, these can be synchronized. It is readily seen (according to C.I.2.2.) that such a space is rigid.

One may thus select a frame in which the point $M \in \sigma$ has constant coordinates that are measurable by rigid segments. It may also be seen that the space σ includes, as a particular case, the inertial space S.

Obvious reasons suggest that these spaces be called spaces whose curvature is constant in time.

Now the following conclusions may be drawn:

I. If in a space one cannot choose a Cartesian frame (or any other frame that is based on the notion of distance) with respect to which all points of space have constant coordinates, i.e., are at rest, that space is non-inertial and non-rigid.

II. If in a space one can find a Cartesian frame (or any other kind that implies distance for coordinates) with respect to which all points of space appear at rest, i.e., their coordinates are constant, then that space is at least rigid and allows synchronization of clocks. In particular such a space is inertial.

III. Any inertial space is rigid. The reciprocal is not necessarily true.

IV. If an observer O, attached to a body C_0 which in turn is bound to a frame Oxyz, notes the variation of shape of C_0 without being aware of any mechanical interaction the body participates in, this means that: (a) the space in which C_0 and O lie is non-inertial and non-rigid; (b) C_0 is in non-inertial motion with respect to the space in which it travels (which may be inertial).

V. If a body C_0 is in non-inertial motion with respect to an inertial space S, but in such a way that the observer bound to C_0 sees space S as rigid (eventually non-inertial but rigid), then one may affirm that C_0 is in a constant deformation motion with respect to S.

VI. If in a frame (a) clocks that may not be synchronized or (b) the synchronized clocks that are placed at the same point of space (so that transmission of information does not differentiate between observations) run at different rhythms, the frame is non-inertial.

The non-inertiality of the frame may originate in its motion (non-inertial, i.e., which does not belong to the same equivalence class as the transmission of information) with respect to an inertial frame or in the non-inertiality of space generated by the presence of masses.

VII. In an inertial space (devoid of a gravitation field) an observer O bound to a body C_0, that is engaged in a non-inertial motion with respect to a frame that is bound to S, watches space S as being S_0 with features of non-inertiality. Such a space S_0, may be termed a non-inertial empty space (empty refers to the absence of a gravitation field). It is exclusively the consequence of non-inertial motion of the frame bound to the body C_0 and of the observer O with respect to the inertial space S.

VIII. The gravitational field is a consequence of the influence of masses. There is no empty space, so that neither is there inertial space, because in any point of the universe the influence of masses is detectable. The inertial space (and the non-inertial empty one) result from an idealization, an approximation which is the closer to reality, the weaker the influence of the gravitational field.

IX. In non-inertial spaces, propagation of information is influenced. This modifies the standard motion, which is just the motion of propagation of information. In turn, the modification of the standard brings about changes in the nature of observations.

It is now the moment to state the general form of the principle of inertia:

In any space, in any domain of space, the motion of any material system undergoes at least the influence of those factors which influence the propagation of information (standard motion), regardless of the selected frame. These factors may be: (a) the gravitational field determined by presence of masses; (b) the non-inertial motion of the frame.

Their influence may be called minimum interaction or non-inertiality effect of the space (or the frame) in question; this influence is identical for all material systems regardless of their rest mass and of whether it originates in the gravitational field generated by the presence of masses or in the non-inertial motion referred to.

Any other interaction but the minimum one represents an exchange of momentum between the material system under study and another material system (or other material systems); this phenomenon is characterized by: (a) the dependence of the interaction effect on the mass (the rest mass as well) of the material system in question and (b) the motion of the information carrier (standard motion) not being altered by interactions which involve exclusively material systems and achieve an exchange of momentum (masses are assumed to be non-zero).

From this formulation of the principle of inertia it follows that:

(1) The principle of equivalence of general relativity is included in this general principle of inertia.

(2) The minimum interaction appears as a characteristic of the space referred to a frame.

(3) If, in particular, the space is inertial (devoid of gravitational field, i.e., the existence of such a field is neglected) and the frame is also inertial, the minimum interaction is nil. It may also appear to be zero, at least within small domains in non-inertial spaces with respect to suitably chosen non-inertial frames.

(4) If the momentum variation of a material system involves, besides the minimum interaction, <u>an exchange of momentum</u>, this means that any change in the motion of system A induced by system B implies an alteration of the motion of B. Thus B detects the opposition against its tendency of changing the state of motion of A which is an essential aspect of inertia. This opposition depends on the mass of A. Therefore the mass measures the inertia and the rest mass measures the intrinsic inertia.

(5) The intrinsic problem of gravitation cannot be considered solved as yet. The theory of general relativity has not solved it either, has not even approached its core, has only looked at it from another point of view.

The gravitational effects are "sensed" by all material systems in the space of the masses that create the field, the information carrier included.

Therefore, the gravitational field was included among the factors that achieve the minimum interaction.

On the other hand, in the case of two masses in gravitational interaction, there is a momentum exchange. It is also true that, for each of the bodies involved in gravitational interaction, the effects are independent of their own mass but dependent on the mass of those bodies with which they interact.

C.II.3. Some aspects of the motion of a particle in a non-inertial situation

As the inertia of a mass particle is expressed in terms of its velocity-dependent mass (according to the known law), it follows that it is impossible to alter, by momentum exchange, the motion of a particle whose velocity approaches that of light, i.e., its inertia is infinite.

In any finite domain of space, the effect of non-inertiality (determined either by gravitational fields or by the non-inertial motion of the frame) appears as minimum when it refers to the information-carrying agent whose speed is large compared to those of material particles. Therefore, in a first approximation, the effect of non-inertiality upon light appears as null and the space as inertial. Thus, the intuitive, usual image of space, which is that of inertial space, is reached.

In the following, the study of non-inertial motion of some particles is outlined in instances that, although particular, are important for the study of nature and for engineering.

C.II.3.1. Accelerated translation. Consider a frame $O_1x_1y_1z_1$ in space S where the gravitational effect of masses is not observed (or rather, neglected). Assume all points $M \in S$ are at rest with respect to $O_1x_1y_1z_1$. Consider also a frame Oxyz that slides along the Ox axis which coincides with O_1x_1; the motion is thus a translation with respect to $O_1x_1y_1z_1$, characterized by the existence of an acceleration which for the observer in $O_1x_1y_1z_1$ appears as equal to a_1 (Fig. C-2.).

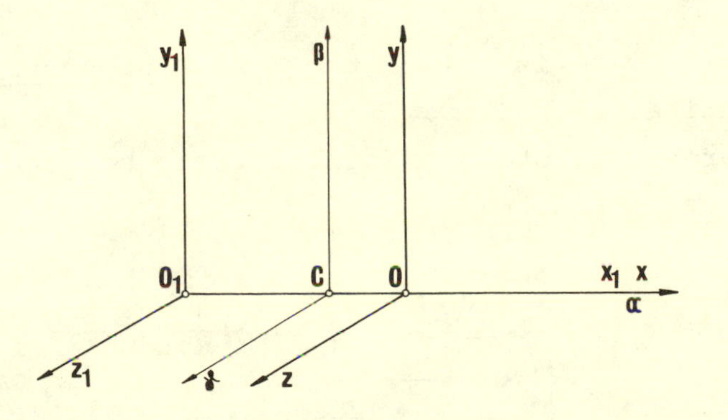

Fig. C-2.

Assume also that at the initial moment, the origins O and O_1 coincide.

To convert the space-time coordinates of one event from the frame $O_1x_1y_1z_1$ to Oxyz (and the reverse), one considers an infinity of frames $C\alpha\beta\gamma$ in inertial motion with respect to $O_1x_1y_1z_1$, so that at a certain time Oxyz coincides with one of the $C\alpha\beta\gamma$ frames and moves at the same velocity as it (with respect to $O_1x_1y_1z_1$).

Obviously, the $C\alpha\beta\gamma$ frames must have (a) suitable velocities and (b) the origins at the initial moment suitably spaced.

The only conceivable way of meeting these conditions

is to see that $C\alpha\beta\gamma$ starts with a velocity $\bar{v}$ relative to Oxyz when $C \equiv O$, i.e., at time $t_1 \equiv t_{01}$ when the space coordinates of both the points O and C in Oxyz are $x_c y_c z_c$; we have

$$x_c = \int_0^{t_{01}} v dt_1, \quad v = \int_0^{t_{01}} a_1 dt_1 = v_c,$$

$$y_c = 0, \quad z_c = 0. \tag{C.45}$$

The $C\alpha\beta\gamma$ frame is called the <u>instantaneous inertial frame</u> identical with Oxyz.

The Lorentz-Einstein transformations between $Ox_1y_1z_1$ and $C\alpha\beta\gamma$ are

$$x_1 - x_c = \frac{\alpha + v_c \tau}{\sqrt{1 - \frac{v_c^2}{c^2}}}, \quad y_1 = \beta, \quad z_1 = \gamma,$$

$$t_1 - t_{01} = \frac{\tau + \frac{v_c}{c^2}}{\sqrt{1 - \frac{v_c^2}{c^2}}}; \tag{C.46}$$

$$\alpha = \frac{x_1 - x_c - v_c (t_1 - t_{01})}{\sqrt{1 - \frac{v_c^2}{c^2}}}, \quad \beta = y_1, \quad \gamma = z_1,$$

$$\tau = \frac{t_1 - t_{01} - \frac{v_c}{c^2} (x_1 - x_c)}{\sqrt{1 - \frac{v_c^2}{c^2}}}, \tag{C.46'}$$

where x_1, y_1, z_1, t_1 are the space-time coordinates in $O_1x_1y_1z_1$ and α, β, γ, τ the coordinates in $C\alpha\beta\gamma$ of the same event E. x, y, z, t are the coordinates of E in Oxyz.

Establishing the relations between x_1, y_1, z_1, t_1 and x, y, z, t is attempted next.

Because frame $C\alpha\beta\gamma$ is inertial and instantaneously identical to Oxyz, one assumes the equations:

$$x = \alpha, \; y = \beta, \; z = \gamma, \; t - t_0 = \tau, \tag{C.47}$$

where t_0 is the lag of time origin arising from the fact that in Oxyz (as in $O_1x_1y_1z_1$) the time origin coincides with the moment when $O \equiv O_1$, while in $C\alpha\beta\gamma$ the origin of time corresponds to $C \equiv O$.

Comparison of eqns. (C.45), (C.46), (C.46') and (C.47) yields:

$$x_1 - x_C = \frac{x + v\,(t - t_0)}{\sqrt{1 - \frac{v^2}{c^2}}}, \; y_1 = y, \; z_1 = z,$$

$$t_1 - t_{01} = \frac{t - t_0 + \frac{v}{c^2}\,x}{\sqrt{1 - \frac{v^2}{c^2}}}; \tag{C.48}$$

$$x = \frac{x_1 - x_C - v\,(t_1 - t_{01})}{\sqrt{1 - \frac{v^2}{c^2}}}, \; y = y_1, \; z = z_1, \tag{C.48'}$$

$$t - t_0 = \frac{t_1 - t_{01} - \frac{v}{c^2}(x_1 - x_c)}{\sqrt{1 - \frac{v^2}{c^2}}}, \qquad \text{(C.48')}$$

which are the transformations sought.

One can readily notice that these relations are similar to those describing the conversion of space-time coordinates between frames Oxyz and $O_1x_1y_1z_1$, were these frames in reciprocally inertial motion. However, taking into account the inner sense of Lorentz-Einstein, transformations, eqns. (C.46), (C.46'), (C.48), (C.48') hold only if:

$$t_1 = t_{01} + dt_1, \; t = t_0 + dt, \; x \in [0, cdt],$$

$$\alpha \in [0, cdt], \; x_1 - x_c = dx_1. \qquad \text{(C.49)}$$

Therefore the transformations above are, in fact, as follows:

$$dx_1 = \frac{dx + vdt}{\sqrt{1 - \frac{v^2}{c^2}}}, \; y_1 = y, \; z_1 = z,$$

$$dt_1 = \frac{dt + \frac{v}{c^2}dx}{\sqrt{1 - \frac{v^2}{c^2}}};$$

$$\text{(C.50)}$$

$$dx = \frac{dx_1 - vdt}{\sqrt{1 - \frac{v^2}{c^2}}}, \; y = y_1, \; z = z_1,$$

$$dt = \frac{dt_1 - \frac{v}{c^2} dx_1}{\sqrt{1 - \frac{v^2}{c^2}}}.$$

Thus, the transformations obtained are valid only throughout an elementary domain in the vicinity of the yOz plane.

If the phenomenon occurs in a certain point of space and requires these transformations, then the yOz plane has to be taken close to the given point.

Eqns. (C.50) and (C.50') lead obviously to transformations of velocities expressed by equations completely similar to eqns. (A.47) and (A.48) that hold for the relative inertial motion of the frames Oxyz and $O_1x_1y_1z_1$. Thus, if a particle moves at velocity $\bar{w}$ with respect to Oxyz, its velocity with respect to $O_1x_1y_1z_1$ is $\bar{w}_1$, and if $\bar{i}$, $\bar{j}$, $\bar{k}$ and $\bar{i}_1$, $\bar{j}_1$, $\bar{k}_1$ are the unit vectors along the Ox, Oy, Oz and O_1x_1, O_1y_1, O_1z_1 axes, respectively, then $\bar{w} = \bar{i}\, w_x + \bar{j}\, w_y + \bar{k}\, w_z$ and $\bar{w}_1 = \bar{i}_1 w_{1x} + \bar{j}_1 w_{1y} + \bar{k}_1 w_{1z}$ and the transformations are:

$$w_x = \frac{w_{1x} - v}{1 - \frac{v}{c^2} w_{1x}}, \quad w_y = \frac{w_{1y} \sqrt{1 - \frac{v^2}{c^2}}}{1 - \frac{v}{c^2} w_{1x}},$$

$$w_z = \frac{w_{1z} \sqrt{1 - \frac{v^2}{c^2}}}{1 - \frac{v}{c^2} w_{1x}}; \qquad (C.51)$$

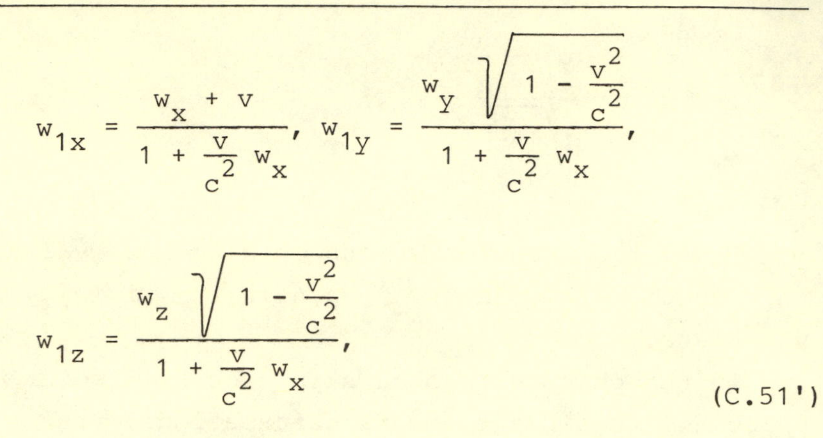

$$w_{1x} = \frac{w_x + v}{1 + \frac{v}{c^2} w_x}, \quad w_{1y} = \frac{w_y \sqrt{1 - \frac{v^2}{c^2}}}{1 + \frac{v}{c^2} w_x},$$

$$w_{1z} = \frac{w_z \sqrt{1 - \frac{v^2}{c^2}}}{1 + \frac{v}{c^2} w_x}, \tag{C.51'}$$

where v is the (variable) relative velocity of Oxyz and $O_1x_1y_1z_1$.

Consider now the well-known differential equation which describes the motion of a particle acted upon by a force:

$$\frac{d}{dt_1} (m_1\bar{w}_1) = \bar{F}_1, \tag{C.52}$$

applied in the inertial frame $O_1x_1y_1z_1$. The rest mass of the particle is:

$$m_1 = \frac{m_0}{\sqrt{1 - \frac{w_1^2}{c^2}}}. \tag{C.53}$$

In Oxyz one may write:

$$\frac{d}{dt} (m\bar{w}) = \bar{F}, \tag{C.54}$$

with

$$m = \frac{m_0}{\sqrt{1 - \frac{w^2}{c^2}}}. \tag{C.55}$$

After routine calculations, eqns. (C.51) and (C.51') yield:

$$w^2 = \frac{(w_1^2 - w_{1x}^2)\left(1 - \frac{v^2}{c^2}\right) + v^2\left(1 - \frac{w_{1x}}{v}\right)^2}{\left(1 - \frac{v}{c^2} w_{1x}\right)^2}, \tag{C.56}$$

$$w_1^2 = \frac{(w^2 - w_x)^2\left(1 - \frac{v^2}{c^2}\right) + v^2\left(1 + \frac{w_x}{v}\right)^2}{\left(1 + \frac{v}{c^2} w_x\right)^2}. \tag{C.56'}$$

Without intending to do a comparative study of the behaviour of the particle of rest mass m_0 in the Oxyz and $O_1x_1y_1z_1$ frames, the discussion is restricted to the case $\bar{F}_1 = 0$. Then, eqns. (C.52) and (C.53) give

$$\frac{m_0\bar{w}_1}{\sqrt{1 - \frac{w_1^2}{c^2}}} = \bar{h}_1 = \text{constant}, \tag{C.57}$$

which implies that $\bar{w}_1 = \bar{w}_{02}$ = const., and, consequently, that the components of $\bar{w}_1$, i.e., w_{1x}, w_{1y}, w_{1z}, are also constant. Thus, $w_{1x} = w_{01x}$ = const.; $w_{1y} = w_{01y}$ = const.; $w_{1z} = w_{01z}$ = const.

Coming now to eqns. (C.51), (C.51') and (C.57) one notices that:

(a) If v = const., i.e., the relative motion of frames is inertial, $w_x = w_{0x}$ = const.; $w_y = w_{0y}$ = const.; $w_z = w_{0z}$ = const. and thus $\bar{w} = \bar{w}_0$ = const.

As a consequence it follows from (C.55) that m = const. and, implicitly, $\bar{F} = 0$.

An identical demonstration establishes the reverse: $\bar{F} = 0$ implies $\bar{F}_1 = 0$, so that it is natural to conclude that, in the relativistic case, if the force is zero in a frame, it is zero in any other frame that is in inertial motion with respect to the former.

(b) If the velocity v is variable (which is the case under consideration), then $\bar{F}_1 = 0$ and, according to (C.57), $\bar{w}_1$ = const. does not imply $\bar{F} = 0$.

Indeed, eqns. (C.51) and (C.56) yield:

$$\bar{F} = \frac{d}{dt}\left\{\frac{m_0\left[\bar{i}(w_{1x}-v) + \sqrt{1-\frac{v^2}{c^2}}\,(\bar{j}w_{1y}+\bar{k}w_{1z})\right]}{\sqrt{1-\frac{1}{c^2}\left[(w_1^2-w_{1x}^2)\left(1-\frac{v^2}{c^2}\right)+(v-w_{1x})^2\right]}}\right\}, \tag{C.58}$$

which is the inertial force in Oxyz, an effect of its motion at variable v with respect to the inertial frame $O_1x_1y_1z_1$.

The effect of this inertial force upon the motion of the particle with respect to Oxyz is independent on the rest mass m_0.

Basically, the phenomenon that gives birth to the inertial force is the following: the particle is subject to no interaction ($\bar{F}_1 = 0$), and the body which is bound to the Oxyz frame is given a non-zero acceleration through interactions, whose result is a variable speed with respect to the $O_1x_1y_1z_1$ inertial frame. Thus, an acceleration field arises in the Oxyz space.

It is noteworthy that, if one assumes $v^2c^{-2} \cong 0$, $w_1^2c^{-2} \cong 0$, $w_{1x}^2c^{-2} \cong 0$, then particularization of eqn. (C.58) yields:

$$\bar{F} = \frac{d}{dt}\{m_0 [\bar{i}(w_{1x} - v) + \bar{j}w_{1y} + \bar{k}w_{1z}]\}, \quad (C.59)$$

i.e., because w_{1x}, w_{1y}, w_{1z} are constant,

$$\bar{F} = -m_0 \frac{dv}{dt}\bar{i} = -\frac{d}{dt}(m_0\bar{v}). \quad (C.59')$$

Thus, neglecting the relativistic effect, one gets to the classical form of the inertial force, obviously in the particular case of the relative translation of reference frames. For the same case, eqn. (C.58) shows that the inertial force in relativistic mechanics depends not only on the derivative of the velocity of relative motion of frames (i.e., on $\dot{v}$) but <u>on the velocity of the particle in the inertial frame $O_1x_1y_1z_1$</u> (i.e., on $\bar{w}_1$) as well. In a very special instance, $w_1 = 0$, eqn. (C.58) becomes:

$$\bar{F} = -\frac{d}{dt}\left[\frac{m_0 v\bar{i}}{\sqrt{1 - \frac{v^2}{c^2}}}\right]. \quad (C.60)$$

<u>Note</u>. These conclusions that refer to the particular case of accelerated translation may be generalized. One can thus reach the expression of inertial force for a complex motion of the Oxyz frame with respect to $O_1x_1y_1z_1$; this motion results from superposition of accelerated translation with rotations that may be accelerated or

uniform. Although this book does not attempt an exhaustive approach but only to present a point of view and suggest possible developments, a generalization regarding these ideas is presented next.

C.II.3.2. <u>A particular gravitational field</u>. Consider the frames $O_1x_1y_1z_1$ and Oxyz (utilized above) in relative accelerated translation motion and suppose that any material particle, whatever its rest mass m_0, is at rest or in uniform rectilinear motion with respect to Oxyz, provided there is no interaction to alter its state of motion with respect to Oxyz frame.

Then $\bar{w} = 0$ or $\bar{w} = \bar{w}_0$ = const. and, consequently, $m_0 (1 - w_0^2 c^{-2})^{-\frac{1}{2}} w_0$ = const., which, according to eqn. (C.54) means $\bar{F} = 0$. Eqns. (C.50'), (C.52) and (C.56') with $\bar{w}_0 = \bar{i}w_{0x} + \bar{j}w_{0y} + \bar{k}w_{0z}$ (and, obviously, w_{0x} = const., w_{0y} = const. and w_{0z} = const.) give:

$$\bar{F}_1 = \frac{d}{dt_1}\left\{\frac{m_0\left[\bar{i}_1(w_{0x}+v) + \sqrt{1-\frac{v^2}{c^2}}\,(\bar{j}_1 w_{0y}+\bar{k}_1 w_{0z})\right]}{1-\frac{1}{c^2}\left[(w_0^2-w_{0x}^2)\left(1-\frac{v^2}{c^2}\right)+(v+w_{0x})^2\right]}\right\} \tag{C.61}$$

the symbols having their known meanings.

For an observer in the $O_1x_1y_1z_1$ frame the force F_1 expressed by eqn. (C.61) is a <u>field force</u> which appears as a <u>gravitational force</u> effect (regarding the motion of the particle with respect to $O_1x_1y_1z_1$) and so does not depend on the rest mass m_0.

However, one can notice that <u>the gravitational force given by (C.61) depends on the velocity of the particle with respect to Oxyz</u>, that is, on w_{0x}, w_{0y}, w_{0z}. Here, in particular, the gravitational force is not an explicit function of the coordinates of the particle, because the discussion is restricted to the particular case of relative rectilinear translation of $O_1x_1y_1z_1$ and Oxyz.

Eqn. (C.61) shows that v = const. implies that the gravitational force is zero ($\bar{F}_1 = 0$), which is quite natural, given the way the particular problem was posed.

Finally, if $\bar{w}_0 = 0$ (i.e., $w_{0x} = w_{0y} = w_{0z} = 0$), eqn. (C.61) becomes:

$$\bar{F}_1 = \frac{d}{dt_1}\left[\frac{m_0 v \bar{i}_1}{\sqrt{1 - \frac{v^2}{c^2}}}\right]. \qquad (C.62)$$

Likewise, assuming that $v^2c^{-2} \cong 0$, $w_0^2c^{-2} \cong 0$, $w_{0x}^2c^{-2} \cong 0$, eqn. (C.61) becomes:

$$F_1 = \frac{d}{dt_1}[m_0(w_{0x} + v)\,\bar{i}_1 + \bar{j}_1 w_{0y} + \bar{k}_1 w_{0z}] =$$

$$m_0 \frac{dv}{dt_1}\,\bar{i}_1, \qquad (C.63)$$

which corresponds to the classical form and which, unlike the relativistic form (C.61), does not depend on the velocity w_0 of the particle but only on the velocity v of $O_1x_1y_1z_1$ and Oxyz with respect to each other.

Note. The similarity of inertial and gravitational forces regarding their quantitative aspect (eqns. (C.58) and (C.61)) and physical effect is evident.

The inertial force acts in inertial systems in

motion, while the gravitational force acts also upon bodies whose motion is referred to frames at rest.

Thus, the inertial force appears as a consequence of (non-inertial) motion of the frame, while the gravitational force which acts upon any particle of the space of a system assumed at rest may not have the same origin, since it is not a consequence of the motion of the frame. Although it is obvious that an absolutely fixed frame is not conceivable, once the gravitational field turns up in a frame this can be considered fixed in calculations, so that the field can not be the effect of frame motion. It then looks as if it is the force of a field which characterizes the space in question (field which is produced by a mass).

There is, however, something conventional in this discussion: a field of forces which alters the state of motion of material particles in a way which is independent of their rest masses has an inertial character (the forces appear as inertial), provided that the frame the motion is referred to is considered to be in non-inertial motion (with respect to another one assumed to be fixed) and that the forces of the field are a consequence of this motion. The forces of the same field may be considered gravitational if the frame is considered fixed. In a system attached to the body C, i.e., fixed with respect to C, the gravitational forces of the mass of the body are also observed.

C.II.3.3. The case of relative motion of the frame. Consider two frames $O_1x_1y_1z_1$, Oxyz. It is assumed that, although these frames may be in non-inertial relative motion (i.e., not necessarily rectilinear uniform translation), the way that their motion (or eventual gravitational fields) influence the propagation of information allows the utilization of Cartesian frames (or any other frames that are based on the notion of distance).

This amounts to admitting, according to the foregoing, that the phenomena under study occur in a rigid space. Let σ_1 be the space of $O_1x_1y_1z_1$ (that is, M_1 $(x_1y_1z_1) \in \sigma_1$ implies x_1 = const., y_1 = const., z_1 = const.) and σ the space of Oxyz (which means that M $(x_1y_1z_1) \in \sigma$ implies x = const., y = const., z = const.).

For any point $M_1 \in \sigma_1$ at any time, there is a point $M \in \sigma$ which coincides with M_1, and conversely. The velocity of M with respect to M_1 is $\bar{v}$, which means that the velocity of M_1 with respect to M is $- \bar{v}$. The vector $\bar{v} = \bar{v}\ (x_1, y_1, z_1, t_1)$ represents the motion of space σ with respect to σ_1, and $\bar{v}' = \bar{v}'\ (x, y, z, t) = - \bar{v}$ the motion of σ_1 with respect to σ.

In the sequel, the following statement is assumed true:

For any elementary space-time domain (i.e., the elementary vicinity of any point in an elementary time interval) the relative motion of spaces σ and σ_1 may be assimilated to an inertial motion of velocity v.

This truth is considered evident: this gives it an axiomatic character and it implies that the relative motion of σ and σ_1 is <u>locally inertial</u>.

Two infinite sets of frames are considered next: the set Γ_1 of frames $M_1\alpha_1\beta_1\gamma_1$ jointly with the space σ_1 whose origins are in all the points $M_1 \in \sigma_1$ and the set Γ of frames $M\alpha\beta\gamma$ jointly with σ, each point $M \in \sigma$ being the origin of such a frame.

Γ_1 and Γ are readily seen to be sets with an infinite number of elements.

Moreover one should specify that these frames are so chosen that axes $M_1\alpha_1$ and $M\alpha$ coincide with each other and with the direction of vector $\bar{v}$ at the point and time where the origins M_1 and M coincide.

The relative motion of the σ and σ_1 spaces, which is generally accelerated, regarded as a locally inertial motion, is thus decomposed into an infinity of inertial motions between frames $M_1\alpha_1\beta_1\gamma_1$ and $M\alpha\beta\gamma$ (Fig. C-3.).

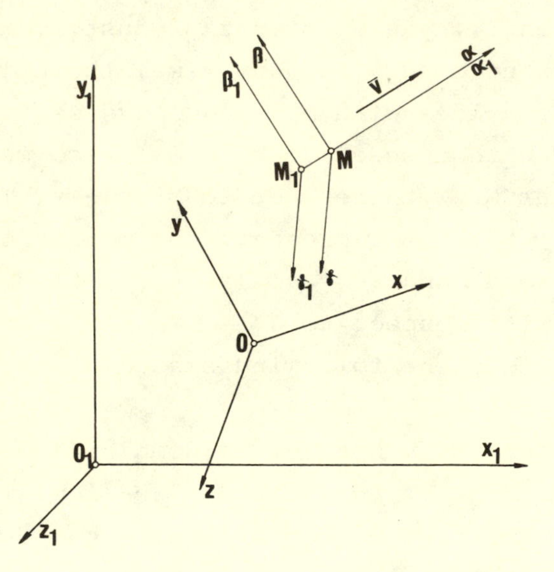

Fig. C-3.

The phenomenon is exactly similar to the approximation to elementary portions of a certain surface by small planar surfaces or of elementary arcs of a curve, by the corresponding chords.

Consider a particle P, which moves both with respect to $O_1x_1y_1z_1$ and Oxyz, the event E of coordinates x_1, y_1, z_1, t_1 in $Ox_1y_1z_1$ and x, y, z, t in Oxyz, representing the passing of P through a certain position, and the event E' of coordinates $x_1 + dx_1$, $y_1 + dy_1$, $z_1 + dz_1$, $t_1 + dt_1$ in $O_1x_1y_1z_1$ and $x + dx$, $y + dy$, $z + dz$, $t + dt$ in Oxyz, representing the passing of the same particle P through an infinitely close position to that characterizing the event E.

Because of the way frames $M_1\alpha_1\beta_1\gamma_1$ and $M\alpha\beta\gamma$ were chosen, the events E and E' have the coordinates: $0,0,0,\tau_1$; $0,0,0,\tau$ and $d\alpha_1$, $d\beta_1$, $d\gamma_1$, $\tau_1 + d\tau_1$; $d\alpha$, $d\beta$, $d\gamma$, $\tau + d\tau$ (with $dt_1 = d\tau_1$ and $dt = d\tau$, since $O_1x_1y_1z_1$ is at rest relative to $M_1\alpha_1\beta_1\gamma_1$ and Oxyz to $M\alpha\beta\gamma$).

The locally inertial character of the relative motion between the frames $M\alpha\beta\gamma$ and $M_1\alpha_1\beta_1\gamma_1$ allows us to write the Lorentz-Einstein transformation relations. They are in this case:

$$d\alpha = \frac{d\alpha_1 - v d\tau_1}{\sqrt{1 - \frac{v^2}{c^2}}}, \quad d\beta = d\beta_1, \quad d\gamma = d\gamma_1,$$

$$d\tau = \frac{d\tau_1 - \frac{v}{c^2} d\alpha_1}{\sqrt{1 - \frac{v^2}{c^2}}}; \tag{C.64}$$

$$d\alpha_1 = \frac{d\alpha + v d\tau}{\sqrt{1 - \frac{v^2}{c^2}}}, \quad d\beta_1 = d\beta, \quad d\gamma_1 = d\gamma,$$

$$d\tau_1 = \frac{d\tau + \frac{v}{c^2} d\alpha}{\sqrt{1 - \frac{v^2}{c^2}}}. \tag{C.64'}$$

Let ϕ_{11}, ϕ_{12}, ϕ_{13} denote the angles between the vector $\bar{v}$ (implicitly, $M_1\alpha_1$ axis) and O_1x_1, O_1y_1 and O_1z_1, respectively, and θ_{11}, θ_{12}, θ_{13} the angles between the same vector (the $M\alpha$ axis) and the axes Ox, Oy, Oz. Likewise let the axes $M_1\beta_1$ and $M_1\gamma_1$ intersect axes O_1x_1, O_1y_1, O_1z_1 at angles ϕ_{21}, ϕ_{22}, ϕ_{23} and ϕ_{31}, ϕ_{32}, ϕ_{33}, respectively. The axes $M\beta$ and $M\gamma$ intersect axes Ox, Oy, Oz at angles θ_{21}, θ_{22}, θ_{23} and θ_{31}, θ_{32}, θ_{33}, respectively.

Then orthogonality yields the equations:

$$\sum_{s=1}^{3} \cos\phi_{is} \cos\phi_{js} = 0,$$

$$\sum_{s=1}^{3} \cos\theta_{is} \cos\theta_{js} = 0;$$

$$i = 1, 2, 3 \text{ and } j = 1, 2, 3. \qquad (C.65)$$

Because $M_1\alpha_1\beta_1\gamma_1$ is at rest with respect to $O_1x_1y_1z_1$, as is $M\alpha\beta\gamma$ with respect to Oxyz, one may write:

$$\begin{aligned} d\alpha_1 &= dx_1 \cos\phi_{11} + dy_1 \cos\phi_{12} + dz_1 \cos\phi_{13}, \\ d\beta_1 &= dx_1 \cos\phi_{21} + dy_1 \cos\phi_{22} + dz_1 \cos\phi_{23}, \\ d\gamma_1 &= dx_1 \cos\phi_{31} + dy_1 \cos\phi_{32} + dz_1 \cos\phi_{33}; \end{aligned} \qquad (C.66)$$

and

$$\begin{aligned} d\alpha &= dx \cos\theta_{11} + dy \cos\theta_{12} + dz \cos\theta_{13}, \\ d\beta &= dx \cos\theta_{21} + dy \cos\theta_{22} + dz \cos\theta_{23}, \\ d\gamma &= dx \cos\theta_{31} + dy \cos\theta_{32} + dz \cos\theta_{33}. \end{aligned} \qquad (C.66')$$

Likewise

$$dx_1 = d\alpha_1 \cos \phi_{11} + d\beta_1 \cos \phi_{21} + d\gamma_1 \cos \phi_{31},$$
$$dy_1 = d\alpha_1 \cos \phi_{12} + d\beta_1 \cos \phi_{22} + d\gamma_1 \cos \phi_{32},$$
$$dz_1 = d\alpha_1 \cos \phi_{13} + d\beta_1 \cos \phi_{23} + d\gamma_1 \cos \phi_{33};$$

(C.67)

and

$$dx = d\alpha \cos \theta_{11} + d\beta \cos \theta_{21} + d\gamma \cos \theta_{31},$$
$$dy = d\alpha \cos \theta_{12} + d\beta \cos \theta_{22} + d\gamma \cos \theta_{32},$$
$$dz = d\alpha \cos \theta_{13} + d\beta \cos \theta_{23} + d\gamma \cos \theta_{33}.$$

(C.67')

Eqns. (C.66) and (C.66') are equivalent to (C.67) and (C.67') by means of eqns. (C.65).

Combining eqns. (C.66) and (C.66') with (C.64) and (C.64'), one gets:

$$dx \cos \theta_{11} + dy \cos \theta_{12} + dz \cos \theta_{13} =$$
$$(dx_1 \cos \phi_{11} + dy_1 \cos \phi_{12} + dz_1 \cos \phi_{13} - vdt_1)$$
$$(1 - v^2c^{-2})^{-\frac{1}{2}},$$
$$dx \cos \theta_{21} + dy \cos \theta_{22} + dz \cos \theta_{23} =$$
$$dx_1 \cos \phi_{21} + dy_1 \cos \phi_{22} + dz_1 \cos \phi_{23},$$
$$dx \cos \theta_{31} + dy \cos \theta_{32} + dz \cos \theta_{33} =$$
$$dx_1 \cos \phi_{31} + dy_1 \cos \phi_{32} + dz_1 \cos \phi_{33},$$
$$dt = [dt_1 - vc^{-2} (dx_1 \cos \phi_{11} + dy_1 \cos \phi_{12} +$$
$$dz_1 \cos \phi_{13})] (1 - v^2c^{-2})^{-\frac{1}{2}};$$

(C.68)

and

$$dx_1 \cos \phi_{11} + dy_1 \cos \phi_{12} + dz_1 \cos \phi_{13} =$$
$$(dx \cos \theta_{11} + dy \cos \theta_{12} + dz \cos \theta_{13} + vdt)$$
$$(1 - v^2c^{-2})^{-\frac{1}{2}},$$

$$dx_1 \cos \phi_{21} + dy_1 \cos \phi_{22} + dz_1 \cos \phi_{23} =$$
$$dx \cos \theta_{21} + dy \cos \theta_{22} + dz \cos \theta_{23},$$
$$dx_1 \cos \phi_{31} + dy_1 \cos \phi_{32} + dz_1 \cos \phi_{33} =$$
$$dx \cos \theta_{31} + dy \cos \theta_{32} + dz \cos \theta_{33},$$
$$dt_1 = [dt + vc^{-2} (dx \cos \theta_{11} + dy \cos \theta_{12} +$$
$$dz \cos \theta_{13})] \cdot (1 - v^2c^{-2})^{-\frac{1}{2}}. \tag{C.68'}$$

These are Lorentz-Einstein type transformations between frames $O_1x_1y_1z_1$ and Oxyz, but they have a <u>local</u> character, since the relative motion of spaces σ and σ_1 was considered <u>locally inertial</u>, i.e., inertial in any elementary space-time domain whose parameters (of direction and velocity) differ continuously from point to point and time to time. As already mentioned, the assimilation of non-inertial relative motion between two spaces σ and σ_1 with a local inertial motion is equivalent to the assimilation of a surface or curve, in an elementary portion, with a plane domain or a line segment, respectively.

Obviously v, ϕ_{ij}, θ_{ij} (i = 1, 2, 3; j = 1, 2, 3) are dependent on time and coordinates.

Consider now: $\bar{i}_1$, $\bar{j}_1$, $\bar{k}_1$; $\bar{i}$, $\bar{j}$, $\bar{k}$; $\bar{u}_1$, $\bar{n}_1$, $\bar{e}_1$; $\bar{u}$, $\bar{n}$, $\bar{e}$ the unit vectors of O_1x_1, O_1y_1, O_1z_1; Ox, Oy, Oz; $M_1\alpha_1$, $M_1\beta_1$, $M_1\gamma_1$ and $M\alpha$, $M\beta$, $M\gamma$ axes, respectively. Obviously $\bar{n} = \bar{n}_1$ and $\bar{e} = \bar{e}_1$ and, at the same time, considering how angles ϕ_{ij} and θ_{ij} (i, j = 1, 2, 3) were defined, one may write:

$$\bar{u}_1 = \bar{i}_1 \cos \phi_{11} + \bar{j}_1 \cos \phi_{12} + \bar{k}_1 \cos \phi_{13},$$
$$\bar{n}_1 = \bar{i}_1 \cos \phi_{21} + \bar{j}_1 \cos \phi_{22} + \bar{k}_1 \cos \phi_{23},$$
$$\bar{e}_1 = \bar{i}_1 \cos \phi_{31} + \bar{j}_1 \cos \phi_{32} + \bar{k}_1 \cos \phi_{33},$$
$$\bar{u} = \bar{i} \cos \theta_{11} + \bar{j} \cos \theta_{12} + \bar{k} \cos \theta_{13},$$

$$\bar{n} = \bar{i} \cos \theta_{21} + \bar{j} \cos \theta_{22} + \bar{k} \cos \theta_{23},$$
$$\bar{e} = \bar{i} \cos \theta_{31} + \bar{j} \cos \theta_{32} + \bar{k} \cos \theta_{33}. \quad (C.69)$$

Eqns. (C.69) together with the relations $d\bar{r}_1 = \bar{i}_1\, dx_1 + \bar{j}_1\, dy_1 + \bar{k}_1\, dz_1$ and $d\bar{r} = \bar{i}\, dx + \bar{j}\, dy + \bar{k}dz$ allow one to rewrite the transformations (C.68) and (C.68') as:

$$\bar{u} \cdot d\bar{r} = \bar{u}_1 \cdot (d\bar{r}_1 - \bar{v}dt_1)\,(1 - v^2c^{-2})^{-\frac{1}{2}},$$
$$\bar{n} \cdot d\bar{r} = \bar{n}_1 \cdot d\bar{r}_1;\ \bar{e} \cdot d\bar{r} = \bar{e}_1 \cdot d\bar{r}_1,$$
$$dt = (dt_1 - vc^{-2}\,\bar{u}_1\, d\bar{r}_1)\,(1 - v^2c^{-2})^{-\frac{1}{2}} \quad (C.70)$$

and

$$\bar{u}_1 \cdot d\bar{r}_1 = \bar{u} \cdot (d\bar{r} + \bar{v}dt)\,(1 - v^2c^{-2})^{-\frac{1}{2}},$$
$$\bar{n}_1 \cdot d\bar{r}_1 = \bar{n} \cdot d\bar{r};\ \bar{e}_1 \cdot d\bar{r}_1 = \bar{e} \cdot d\bar{r},$$
$$dt_1 = (dt + vc^{-2}\,\bar{u}\, d\bar{r})\,(1 - v^2c^{-2})^{-\frac{1}{2}}, \quad (C.70')$$

respectively.

The velocity of P, whose motion was studied above, is obviously: $\bar{w}_1 = \frac{d\bar{r}_1}{dt_1}$ in $O_1x_1y_1$ (i.e., in $M_1\alpha_1\beta_1\gamma_1$ as well) and $\bar{w} = \frac{d\bar{r}}{dt}$ in Oxyz (and $M\alpha\beta\gamma$ as well).

Eqns. (C.70) and (C.70') yield, after transformations:

$$\bar{u} \cdot \bar{w} = \frac{\bar{u}_1 \cdot (\bar{w}_1 - \bar{v})}{1 - \frac{v}{c^2}\,\bar{u}_1 \cdot \bar{w}_1};$$

$$\bar{n} \cdot \bar{w} = \frac{\bar{n}_1 \cdot \bar{w}_1 \sqrt{1 - \frac{v^2}{c^2}}}{1 - \frac{v}{c^2}\,\bar{u}_1 \cdot \bar{w}_1};$$

$$\bar{e} \cdot \bar{w} = \frac{\bar{e}_1 \cdot \bar{w}_1 \sqrt{1 - \frac{v^2}{c^2}}}{1 - \frac{v}{c^2} \bar{u}_1 \cdot \bar{w}_1}, \qquad (C.71)$$

and

$$\bar{u}_1 \cdot \bar{w}_1 = \frac{\bar{u} \cdot (\bar{w} + \bar{v})}{1 + \frac{v}{c^2} \bar{u} \cdot \bar{w}};$$

$$\bar{n}_1 \cdot \bar{w}_1 = \frac{\bar{n} \cdot \bar{w} \sqrt{1 - \frac{v^2}{c^2}}}{1 + \frac{v}{c^2} \bar{u} \cdot \bar{w}};$$

$$\bar{e}_1 \cdot \bar{w}_1 = \frac{\bar{e} \cdot \bar{w} \sqrt{1 - \frac{v^2}{c^2}}}{1 + \frac{v}{c^2} \bar{u} \cdot \bar{w}}. \qquad (C.71')$$

On the other hand $w^2 = (\bar{u} \cdot \bar{w})^2 + (\bar{n} \cdot \bar{w})^2 + (\bar{e} \cdot \bar{w})^2$ and $w_1^2 = (\bar{u}_1 \cdot \bar{w}_1)^2 + (\bar{n}_1 \cdot \bar{w}_1)^2 + (\bar{e}_1 \cdot \bar{w}_1)^2$.

It turns out that, among the consequences of eqns. (C.71) and (C.71'), is the equation:

$$w^2 = \frac{\left[w_1^2 - (\bar{u}_1 \cdot \bar{w}_1)^2\right]\left(1 - \frac{v^2}{c^2}\right) + v^2\left(1 - \frac{\bar{u}_1 \cdot \bar{w}_1}{v}\right)^2}{\left(1 - \frac{v}{c^2} \bar{u}_1 \cdot \bar{w}_1\right)^2}, \qquad (C.72)$$

or, assuming that the velocity of M with respect to $O_1x_1y_1z_1$ is $\bar{v} = \bar{u}_1 v$ and the velocity of M_1 with respect to Oxyz is $\bar{v}_1 = -\bar{v} = -v\bar{u}$, it follows that:

$$w^2 = \frac{w_1^2 \left[1 - \frac{(\bar{v} \cdot \bar{w}_1)^2}{v^2 w_1^2} \right] \left(1 - \frac{v^2}{c^2} \right) + v^2 \left(1 - \frac{\bar{v} \cdot \bar{w}_1}{v^2} \right)^2}{\left(1 - \frac{\bar{v} \cdot \bar{w}_1}{c^2} \right)^2}; \tag{C.73}$$

and similarly:

$$w_1^2 = \frac{w^2 \left[1 - \frac{(\bar{v} \cdot \bar{w})^2}{v^2 w^2} \right] \left(1 - \frac{v^2}{c^2} \right) + v^2 \left[1 + \frac{(\bar{v} \cdot \bar{w})}{w^2} \right]^2}{\left(1 + \frac{\bar{v} \cdot \bar{w}}{c^2} \right)^2}. \tag{C.73'}$$

It is known that the differential equation of the motion of a particle in $O_1x_1y_1z_1$ is:

$$\frac{d}{dt_1} \left(\frac{m_0 \bar{w}_1}{\sqrt{1 - \frac{w_1^2}{c^2}}} \right) = \bar{F}_1, \tag{C.74}$$

and in Oxyz

$$\frac{d}{dt} \left(\frac{m_0 \bar{w}}{\sqrt{1 - \frac{w^2}{c^2}}} \right) = \bar{F}, \tag{C.74'}$$

the rest mass m_0 having an invariant character.

If the frame $O_1x_1y_1z_1$ is considered the main one, i.e., all motions are referred to it (it may, inadequately, be called the fixed frame), then the particular case

$\bar{F}_1 = 0$ implies w_1 = const.; $\bar{F} \neq 0$ is then an inertial force and $\bar{F} = 0$ implies $\bar{w}$ = const., which may confer on the vector $\bar{F}_1 \neq 0$ the character of a gravitational force.

An improved form may be given to these ideas if one associates eqns. (C.69) with (C.71), (C.71'), (C.73) and (C.73') and takes into account the equations:

$$\begin{aligned} w_\alpha &= \bar{u} \cdot \bar{w} = w_x \cos\theta_{11} + w_y \cos\theta_{12} + w_z \cos\theta_{13}, \\ w_\beta &= \bar{n} \cdot \bar{w} = w_x \cos\theta_{21} + w_y \cos\theta_{22} + w_z \cos\theta_{23}, \\ w_\gamma &= \bar{e} \cdot \bar{w} = w_x \cos\theta_{31} + w_y \cos\theta_{32} + w_z \cos\theta_{33}; \end{aligned} \tag{C.75}$$

with $\bar{w} = \bar{i}\, w_x + \bar{j}\, w_y + \bar{k}\, w_z$, the analytical expression of the vector $\bar{w}$ in Oxyz and with:

$$\begin{aligned} w_{\alpha 1} &= \bar{u}_1 \cdot \bar{w}_1 = w_{1x} \cos\phi_{11} + w_{1y} \cos\phi_{12} + w_{1z} \cos\phi_{13}, \\ w_{\beta 1} &= \bar{n}_1 \cdot \bar{w}_1 = w_{1x} \cos\phi_{21} + w_{1y} \cos\phi_{22} + w_{1z} \cos\phi_{23}, \\ w_{\gamma 1} &= \bar{e}_1 \cdot \bar{w}_1 = w_{1x} \cos\phi_{31} + w_{1y} \cos\phi_{32} + w_{1z} \cos\phi_{33}; \end{aligned} \tag{C.75'}$$

where $\bar{w}_1 = \bar{i}_1\, w_{1x} + \bar{j}_1\, w_{1y} + \bar{k}_1\, w_{1z}$ is the analytical expression of $\bar{w}_1$ in the frame $O_1x_1y_1z_1$.

Thus eqns. (C.71) and (C.71') become

$$w_\alpha = \frac{w_{\alpha 1} - v}{1 - \frac{v}{c^2} w_{\alpha 1}}; \quad w_\beta = \frac{w_{\beta 1} \sqrt{1 - \frac{v^2}{c^2}}}{1 - \frac{v}{c^2} w_{\alpha 1}};$$

$$w_\gamma = \frac{w_{\gamma 1} \sqrt{1 - \frac{v^2}{c^2}}}{1 - \frac{v}{c^2} w_{\alpha 1}}; \tag{C.76}$$

or

$$w_x \cos \theta_{11} + w_y \cos \theta_{12} + w_z \cos \theta_{13} =$$
$$(w_{1x} \cos \phi_{11} + w_{1y} \cos \phi_{12} + w_{1z} \cos \phi_{13} - v)$$
$$[1 - vc^{-2} (w_{1x} \cos \phi_{11} + w_{1y} \cos \phi_{12} +$$
$$w_{1z} \cos \phi_{13})]^{-1},$$
$$w_x \cos \theta_{21} + w_y \cos \theta_{22} + w_z \cos \theta_{23} =$$
$$(w_{1x} \cos \phi_{21} + w_{1y} \cos \phi_{22} + w_{1z} \cos \phi_{23})$$
$$(1 - v^2c^{-2})^{\frac{1}{2}}]\ [1 - vc^{-2} (w_{1x} \cos \phi_{11} +$$
$$w_{1y} \cos \phi_{12} + w_{1z} \cos \phi_{13})]^{-1},$$
$$w_x \cos \theta_{31} + w_y \cos \theta_{32} + w_z \cos \theta_{33} =$$
$$(w_{1x} \cos \phi_{31} + w_{1y} \cos \phi_{32} + w_{1z} \cos \phi_{33})$$
$$(1 - v^2c^{-2})^{\frac{1}{2}}\ [1 - vc^2 (w_{1x} \cos \phi_{11} +$$
$$w_{1y} \cos \phi_{12} + w_{1z} \cos \phi_{13})]^{-1}, \tag{C.77}$$

and

$$w_{\alpha 1} = \frac{w_\alpha + v}{1 + \frac{v}{c^2} w_\alpha}; \quad w_{\beta 1} = \frac{w_\beta \sqrt{1 - \frac{v^2}{c^2}}}{1 + \frac{v}{c^2} w_\alpha};$$

$$w_{\gamma 1} = \frac{w_\gamma \sqrt{1 - \frac{v^2}{c^2}}}{1 + \frac{v}{c^2} w_\alpha}; \tag{C.78}$$

or

$$w_{1x} \cos \phi_{11} + w_{1y} \cos \phi_{12} + w_{1z} \cos \phi_{13} =$$
$$(w_x \cos \theta_{11} + w_y \cos \theta_{12} + w_z \cos \theta_{13} + v)$$
$$[1 + vc^{-2} (w_x \cos \theta_{11} + w_y \cos \theta_{12} +$$
$$w_z \cos \theta_{13})]^{-1},$$

$$w_{1x} \cos \phi_{21} + w_{1y} \cos \phi_{22} + w_{1z} \cos \phi_{23} = (w_x \cos \theta_{21} + w_y \cos \theta_{22} + w_z \cos \theta_{23}) (1 - v^2c^{-2})^{\frac{1}{2}} [1 + vc^{-2} (w_x \cos \theta_{11} + w_y \cos \theta_{12} + w_z \cos \theta_{13})]^{-1},$$

$$w_{1x} \cos \phi_{31} + w_{1y} \cos \phi_{32} + w_{1z} \cos \phi_{33} = (w_x \cos \theta_{31} + w_y \cos \theta_{32} + w_z \cos \theta_{33}) (1 - v^2c^{-2})^{\frac{1}{2}} [1 + vc^{-2} (w_x \cos \theta_{11} + w_y \cos \theta_{12} + w_z \cos \theta_{13})]^{-1}. \quad (C.79)$$

Under these circumstances (i.e., on the grounds of eqns. (C.75), (C.75'), (C.77) and (C.79), eqns. (C.74) and (C.74') may be projected on the $M_1\alpha_1$, $M_1\beta_1$, $M_1\gamma_1$ axes and the $M\alpha$, $M\beta$, $M\gamma$ axes; if the variations of $\bar{u}_1$, $\bar{n}_1$, $\bar{e}_1$ and $\bar{u}$, $\bar{n}$, $\bar{e}$ unit vectors are neglected, one gets:

$$F_{1\alpha 1} = F'_{1x_1} \cos \phi_{11} + F'_{1y_1} \cos \phi_{12} + F'_{1z_1} \cos \phi_{13} = \frac{d}{dt_1} \frac{m_0 w_{\alpha 1}}{\sqrt{1 - \frac{w_1^2}{c^2}}} =$$

$$\frac{d}{dt_1} \{m_0 (w_x \cos \theta_{11} + w_y \cos \theta_{12} + w_z \cos \theta_{13} + v) [1 + vc^{-2} (w_x \cos \theta_{11} + w_y \cos \theta_{12} + w_z \cos \theta_{13})]^{-1} [1 - [w^2 (1 - (\bar{v} \cdot \bar{w})^2 v^{-2} w^{-2}) (1 - v^2c^{-2}) + v^2 (1 + \bar{v} \cdot \bar{w}v^{-2})^2] c^{-2} (1 + \bar{v} \cdot \bar{w}c^{-2})^{-2}]^{-\frac{1}{2}}\},$$

$$F_{1\beta 1} = F'_{1x_1} \cos \phi_{21} + F'_{1y_1} \cos \phi_{22} + F'_{1z_1} \cos \phi_{23} = \frac{d}{dt_1} \frac{m_0 w_{\beta 1}}{\sqrt{1 - \frac{w_1^2}{c^2}}} =$$

$$\frac{d}{dt_1} \{m_0 (w_x \cos \theta_{21} + w_y \cos \theta_{22} + w_z \cos \theta_{23}) (1 - v^2c^{-2})^{\frac{1}{2}} [1 + vc^{-2} (w_x \cos \theta_{11} + w_y \cos \theta_{12} +$$

$$w_z \cos \theta_{13})] [1 - [w^2 (1 - (\bar{v} \cdot \bar{w})^2 v^{-2} w^{-2}) (1 - v^2c^{-2}) + v^2 (1 + \bar{v} \cdot \bar{w}v^{-2})^2]c^{-2} (1 + \bar{v} \cdot \bar{w}c^{-2})^{-2}]^{-\frac{1}{2}}\},$$

$$F_{1\gamma 1} = F'_{1x_1} \cos \phi_{31} + F'_{1y_1} \cos \phi_{32} + F'_{1z_1} \cos \phi_{33} = \frac{d}{dt_1} \frac{m_0 w_{\gamma 1}}{\sqrt{1 - \frac{w_1^2}{c^2}}} =$$

$$\frac{d}{dt_1} \{m_0 (w_x \cos \theta_{31} + w_y \cos \theta_{32} + w_z \cos \theta_{33}) (1 - v^2c^{-2})^{\frac{1}{2}} [1 + vc^{-2} (w_x \cos \theta_{11} + w_y \cos\theta_{12} + w_z \cos \theta_{13})]^{-1} [1 - [w^2 (1 - (\bar{v} \cdot \bar{w})^2 v^{-2} w^{-2}) (1 - v^2c^{-2}) + v^2 (1 + \bar{v} \cdot \bar{w}v^{-2})^2]c^{-2} (1 + \bar{v} \cdot \bar{w}c^{-2})^{-2}]^{-\frac{1}{2}}\} \tag{C.80}$$

and

$$F_\alpha = F'_x \cos \theta_{11} + F'_y \cos \theta_{12} + F'_z \cos \theta_{13} = \frac{d}{dt} \frac{m_0 w_\alpha}{\sqrt{1 - \frac{w^2}{c^2}}} = \frac{d}{dt} \{m_0 (w_{1x} \cos \phi_{11} + w_{1y} \cos \phi_{12} + w_{1z} \cos \phi_{13} - v) [1 - vc^{-2} (w_{1x} \cos \phi_{11} + w_{1y} \cos \phi_{12} + w_{1z} \cos \phi_{13})^{-1} [1 - [w_1^2 (1 - (\bar{v} \cdot \bar{w}_1)^2 v^{-2} w_1^{-2}) (1 - v^2c^{-2}) + v^2 (1 - \bar{v} \cdot \bar{w}_1v^{-2})^2]c^{-2} (1 - \bar{v} \cdot \bar{w}_1c^{-2})^{-2}]^{-\frac{1}{2}}\},$$

$$F_\beta = F'_x \cos \theta_{21} + F'_y \cos \theta_{22} + F'_z \cos \theta_{23} = \frac{d}{dt} \frac{m_0 w_\beta}{\sqrt{1 - \frac{w^2}{c^2}}} = \frac{d}{dt} \{m_0 (w_{1x} \cos \phi_{21} + w_{1y} \cos \phi_{22} + w_{1z} \cos \phi_{23}) (1 - v^2c^{-2})^{\frac{1}{2}} [1 - vc^{-2} (w_{1x} \cos \phi_{11} + w_{1y} \cos \phi_{12} + w_{1z} \cos \phi_{13})]^{-1} [1 - [w_1^2 (1 - (\bar{v} \cdot \bar{w}_1)^2 v^{-2}w_1^{-2}) (1 - v^2c^{-2}) + v^2 (1 - \bar{v} \cdot \bar{w}_1v^{-2})^2]c^{-2} (1 - \bar{v} \cdot \bar{w}_1c^{-2})^{-2}]^{-\frac{1}{2}}\},$$

$$F_\gamma = F'_x \cos\theta_{31} + F'_y \cos\theta_{32} + F'_z \cos\theta_{33} =$$

$$\frac{d}{dt}\frac{m_0 w_\gamma}{\sqrt{1-\frac{w^2}{c^2}}} = \frac{d}{dt}\{m_0\,(w_{1x}\cos\phi_{31} +$$

$$w_{1y}\cos\phi_{32} + w_{1z}\cos\phi_{33})\,(1 - v^2c^{-2})^{\frac{1}{2}}$$
$$[1 - vc^{-2}\,(w_{1x}\cos\phi_{11} + w_{1y}\cos\phi_{12} +$$
$$w_{1z}\cos\phi_{13})]^{-1}\,[1 - [w_1^2\,(1 - (\bar{v}\cdot\bar{w}_1)^2\,v^{-2}w_1^{-2})$$
$$(1 - v^2c^{-2}) + (1 - \bar{v}\cdot\bar{w}_1 v^{-2})^2\,v^2c^{-2}$$
$$(1 - \bar{v}\cdot\bar{w}_1 c^{-2})^{-2}]^{-\frac{1}{2}}\}. \qquad (C.81)$$

However in (C.74') one gets $\bar{w} = \bar{u}\,\bar{w}_\alpha + \bar{n}\,\bar{w}_\beta + \bar{e}\,\bar{w}_\gamma$ which means that

$$\bar{F} = \frac{d}{dt}\frac{m_0\bar{w}}{\sqrt{1-\frac{w^2}{c^2}}} = \frac{d}{dt}\left(m_0\,\frac{\bar{u}w_\alpha + \bar{n}w_\beta + \bar{e}w_\gamma}{\sqrt{1-\frac{w^2}{c^2}}}\right) =$$

$$\bar{u}F_\alpha + \bar{n}F_\beta + \bar{e}F_\gamma + \frac{m_0}{\sqrt{1-\frac{w^2}{c^2}}}$$

$$\left(w_\alpha\frac{d\bar{u}}{dt} + w_\beta\frac{d\bar{n}}{dt} + w_\gamma\frac{d\bar{e}}{dt}\right), \qquad (C.82)$$

because the unit vectors $\bar{u}$, $\bar{n}$, $\bar{e}$ are variable in direction, i.e., their set is in correspondence with the set of moments.

Similarly, eqn. (C.74) has the following meaning:

$$\bar{F}_1 = \frac{d}{dt_1}\frac{m_0\bar{w}_1}{\sqrt{1-\frac{w_1^2}{c^2}}} =$$

$$\frac{d}{dt_1}\left(m_0 \frac{\bar{u}_1 w_{\alpha 1} + \bar{n}_1 w_{\beta 1} + \bar{e}_1 w_{\gamma 1}}{\sqrt{1 - \frac{w_1^2}{c^2}}} \right) =$$

$$\bar{u}_1 F_{1\alpha_1} + \bar{n}_1 F_{1\beta_1} + \bar{e}_1 F_{1\gamma_1} + \frac{m_0}{\sqrt{1 - \frac{w_1^2}{c^2}}}$$

$$\left(w_{\alpha 1} \frac{d\bar{u}_1}{dt_1} + w_{\beta 1} \frac{d\bar{n}_1}{dt_1} + w_{\gamma 1} \frac{d\bar{e}_1}{dt_1} \right) \tag{C.83}$$

because one has to consider the variation of the unit vectors $\bar{u}_1$, $\bar{n}_1$, $\bar{e}_1$ generated by the correspondence between their set and that of momenta.

Eqns. (C.69) have obviously to be considered in calculations.

Eqns. (C.80) and (C.81) have evidently taken account of eqns. (C.73) and (C.73').

The functions:

$$\phi_{ij} = \phi_{ij}\ (x_1,\ y_1,\ z_1,\ t_1)\ \text{and}$$

$$\theta_{ij} = \theta_{ij}\ (x,\ y,\ z,\ t);$$

$$i,j = 1,\ 2,\ 3 \tag{C.84}$$

are known.

Under these circumstances, eqns. (C.80), (C.81), (C.82) and (C.83) may be used to calculate (by elementary means) the components of the forces $\bar{F}$ in Oxyz and $\bar{F}_1$ in $O_1x_1y_1z_1$, components, which are not the same as F'_x, F'_y, F'_z, F'_{1x_1}, F'_{1y_1}, F'_{1z_1}, except for the case of constant $\bar{u}$, $\bar{n}$, $\bar{e}$, $\bar{u}_1$, $\bar{n}_1$, $\bar{e}_1$ unit vectors.

Considering also the relations of transformation of space-time coordinates (C.68) and (C.68') for the case when

$$\begin{aligned} w_{1x} &= w_{01x} = \text{constant}, \\ w_{1y} &= w_{01y} = \text{constant}, \\ w_{1z} &= w_{01z} = \text{constant}, \end{aligned} \tag{C.85}$$

the force $\bar{F}$ appears as an inertial force and if:

$$\begin{aligned} w_{x} &= w_{0x} = \text{constant}, \\ w_{y} &= w_{0y} = \text{constant}, \\ w_{z} &= w_{0z} = \text{constant}, \end{aligned} \tag{C.86}$$

$\bar{F}_1$ is gravitational (because it was assumed that frame $O_1x_1y_1z_1$ was fixed, i.e., was a frame to which motions are ultimately referred).

Obviously these considerations and calculations are suitable for extensive developments; however this section is intended only to suggest the approach.

A more general treatment would include the previous case, of translation, as a particular case.

Two simple cases are considered next:

(1) Let the frame $O_1x_1y_1z_1$ be assumed fixed and the frame Oxyz in uniform rotation (about the $Oy = O_1y_1$ common axis) with respect to $O_1x_1y_1z_1$, as in Fig. C-4.

Assume $\psi = \omega t$, where ω is a constant.

Consider the point M_1 fixed with respect to $O_1x_1y_1z_1$ frame; now the (inertial) force, which, for an observer attached to Oxyz, seems to act upon M_1 whose rest mass is m_0, may be calculated.

With the notations specified during the general

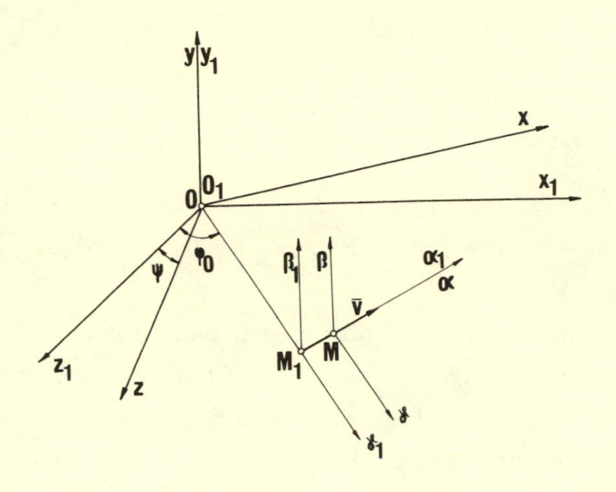

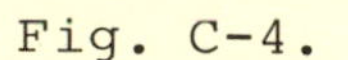
Fig. C-4.

treatment and according to Fig. C-4 one may write:

$$\phi_{11} = \phi_0, \quad \phi_{12} = \frac{\pi}{2}, \quad \phi_{13} = \phi_0 + \frac{\pi}{2},$$

$$\phi_{21} = \frac{\pi}{2}, \quad \phi_{22} = 0, \quad \phi_{23} = \frac{\pi}{2},$$

$$\phi_{31} = \frac{\pi}{2} - \phi_0, \quad \phi_{32} = \frac{\pi}{2}, \quad \phi_{33} = \phi_0,$$

$$\theta_{11} = \phi_0 - \omega t, \quad \theta_{12} = \frac{\pi}{2}, \quad \theta_{13} = \frac{\pi}{2} + \phi_0 - \omega t,$$

$$\theta_{21} = \frac{\pi}{2}, \quad \theta_{22} = 0, \quad \theta_{23} = \frac{\pi}{2},$$

$$\theta_{31} = \frac{\pi}{2} - \phi_0 + \omega t, \quad \theta_{32} = \frac{\pi}{2}, \quad \theta_{33} = \phi_0 - \omega t.$$

Meanwhile, since M_1 is fixed with respect to $O_1x_1y_1z_1$, in this particular case:

$$w_{1x} = w_{1y} = w_{1z} = 0, \quad w_{\alpha 1} = w_{\beta 1} = w_{\gamma 1} = 0.$$

Replacement of the formulas above and consideration of the data of the problem yield:

$$\bar{u} = \bar{i} \cos (\phi_0 - \omega t) - \bar{k} \sin (\phi_0 - \omega t),$$
$$\bar{n} = \bar{j},$$
$$\bar{e} = \bar{i} \sin (\phi_0 - \omega t) + \bar{k} \sin (\phi_0 - \omega t),$$
$$v = R\omega, \ w_\alpha = - R\omega, \ w_\beta = 0, \ w_\gamma = 0.$$

It is noteworthy that derivation with respect to time in eqns. (C.82) must be effected assuming the unit vectors $\bar{i}$, $\bar{j}$, $\bar{k}$ of the frame Oxyz constant (although the axes rotate with respect to $O_1x_1y_1z_1$) because the variations are noted only by an observer that is bound to Oxyz frame.

One can see that:

$$F_\alpha = F'_x \cos (\phi_0 - \omega t) - F'_z \sin (\phi_0 - \omega t) =$$

$$\frac{d}{dt}\left(- \frac{m_0 R\omega}{\sqrt{1 - \frac{R^2\omega^2}{c^2}}}\right)$$

and because ω = const., $F_x = 0$.

Eqns. (C.81) and (C.82) yield immediately:

$$F_\beta = F_y = 0, \ F_\gamma = F'_x \sin (\phi_0 - \omega t) +$$

$$F'_z \cos (\phi_0 - \omega t) = 0,$$

since $w_\gamma = 0$, and

$$w_\alpha \frac{d\bar{u}}{dt} + w_\beta \frac{d\bar{n}}{dt} + w_\gamma \frac{d\bar{e}}{dt} = w_\alpha \frac{d\bar{u}}{dt} =$$

$$= - R\omega \frac{d}{dt} [\bar{i} \cos (\phi_0 - \omega t) - \bar{k} \sin (\phi_0 - \omega t)] =$$

$$- R\omega^2 [\bar{i} \sin (\phi_0 - \omega t) + \bar{k} \cos (\phi_0 - \omega t)] =$$

$$- R\omega^2 \bar{e}$$

or

$$\bar{F} = - \frac{m_0 R\omega^2 \bar{e}}{\sqrt{1 - \frac{R^2\omega^2}{c^2}}},$$

which is the inertial force sought.

It is worth noting that, for $R^2\omega^2 c^{-2} \cong 0$, the equation known in classical mechanics for the case in question is obtained.

(2) Now let two frames $O_1x_1y_1z_1$ and Oxyz be in relative rectilinear uniformly accelerated motion along the axes O_1z_1 and Oz superposed as in Fig. C-5. Let a_0 denote the acceleration (whose sense is opposite to that

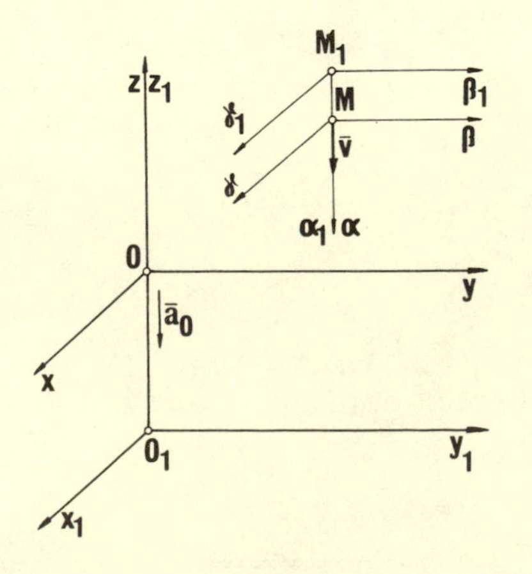

Fig. C-5.

of the Oz and O_1z_1 axes) for an observer who participates in the motion of the frame $O_1x_1y_1z_1$ to which all motions are referred (i.e., which is assumed fixed). Now the gravitational field noted by the observer to act upon a particle M at rest with respect to Oxyz is sought; the rest mass of the particle M is m_0. One can write:

$$\phi_{11} = \frac{\pi}{2},\ \phi_{12} = \frac{\pi}{2},\ \phi_{13} = \pi,\ \phi_{21} = \frac{\pi}{2},\ \phi_{22} = 0,$$

$$\phi_{23} = \frac{\pi}{2},\ \phi_{31} = 0,\ \phi_{32} = \frac{\pi}{2},\ \phi_{33} = \frac{\pi}{2};$$

$$\theta_{11} = \frac{\pi}{2},\ \theta_{12} = \frac{\pi}{2},\ \theta_{13} = \pi,\ \theta_{21} = \frac{\pi}{2},\ \theta_{22} = 0,$$

$$\theta_{23} = \frac{\pi}{2},\ \theta_{31} = 0,\ \theta_{32} = \frac{\pi}{2},\ \theta_{33} = \frac{\pi}{2};$$

$$w_{x1} = w_y = w_z = 0,\quad w_\alpha = w_\beta = w_\gamma = 0,\ w = 0,$$

$$w_{\alpha 1} = v,\ w_{\beta 1} = 0,\ w_{\gamma 1} = 0,$$

$$w_{1z} = -v,\ w_{1y} = 0,\ w_{1z} = 0,$$

$$\bar{u}_1 = -\bar{k}_1,\ \bar{n}_1 = \bar{j}_1,\ \bar{e}_1 = \bar{i}_1,\ \bar{v} = a_0t_1\bar{u}_1 = -a_0t_1\bar{k}_1,$$

$$\frac{d\bar{u}_1}{dt_1} = 0,\ \frac{d\bar{n}_1}{dt_1} = 0,\ \frac{d\bar{e}_1}{dt_1} = 0,$$

which means that $F_{1x_1} = F'_{1x_1}$, $F_{1y_1} = F'_{1y_1}$, and $F_{1z_1} = F'_{1z_1}$ with the notations employed above.

Eqns. (C.80) gives:

$$F_{1\alpha_1} = -F_{1z_1} = \frac{m_0a_0}{\sqrt{1 - \frac{v^2}{c^2}}} = \frac{m_0a_0}{\sqrt{1 - \frac{a_0^2t_1^2}{c^2}}},$$

$$F_{1\beta_1} = F_{1y_1} = 0, \; F_{1\gamma_1} = F_{1x_1} = 0,$$

so that eqn. (C.83) becomes:

$$\bar{F}_1 = \bar{u}_1 \frac{m_0 a_0}{\sqrt{1 - \frac{a_0^2 t_1^2}{c^2}}} = -\bar{k}_1 \frac{m_0 a_0}{\sqrt{1 - \frac{a_0^2 t_1^2}{c^2}}}.$$

It is readily seen that for the (particular) case $a_0^2 t_1^2/c^2 \cong 0$, one reaches the trivial equation $F_1 = - k_1 m_0 a_0$, well known in Newtonian mechanics. However, the distinction between the equations presented now and the similar ones in classical mechanics does not consist only in a factor of

$$\left(1 - \frac{a_0^2 t_1^2}{c^2}\right)^{-\frac{1}{2}}.$$

Indeed, now one admits that the velocity of particle M with respect to Oxyz is constant, $\bar{w} = b_1 \bar{i} + b_2 \bar{j} + b_3 \bar{k}$, where b_1, b_2, b_3 are given constants. Then eqn. (C.80) allows one to conclude:

$$F_{1\alpha_1} = - F_{1z_1} = \frac{d}{dt_1}\left\{ m_0 \, (-b_3 + a_0 t_1) \left[1 + \frac{a_0 t_1}{c^2}(-b_3)\right]^{-1} \left[1 - \frac{a_0^2 t_1^2}{c^2}\left(1 - \frac{b_3}{a_0 t_1}\right)^2 \left(1 - \frac{a_0 b_3 t_1}{c^2}\right)^{-2}\right]^{\frac{1}{2}}\right\},$$

$$F_{1\beta_1} = \frac{d}{dt_1}\left\{ m_0 b_2 \left(1 - \frac{a_0^2 t_1^2}{c^2}\right)^{\frac{1}{2}} \left[1 + \frac{a_0 t_1}{c^2}(-b_3)\right]^{-1} \left[1 - \frac{a_0^2 t_1^2}{c^2}\left(1 - \frac{b_3}{a_0 t_1}\right)^2 \left(1 - \frac{a_0 b_3 t_1}{c^2}\right)^{-2}\right]^{-\frac{1}{2}}\right\},$$

$$F_{1\gamma_1} = \frac{d}{dt_1}\left\{ m_0 b_1 \left(1 - \frac{a_0^2 t_1^2}{c^2}\right)^{\frac{1}{2}} \left[1 + \frac{a_0 t_1}{c^2}(-b_3)\right]^{-1} \left[1 - \frac{a_0^2 t_1^2}{c^2}\left(1 - \frac{b_3}{a_0 t_1}\right)^2 \left(1 - \frac{a_0 b_3 t_1}{c^2}\right)^{-2}\right]^{-\frac{1}{2}}\right\},$$

so that after elementary calculations one can write:

$$F_{1\alpha_1} = \frac{d}{dt_1} \frac{m_0 \, (a_0 t_1 - b_3)}{\sqrt{1 - \frac{b_3^2}{c^2}}\sqrt{1 - \frac{a_0^2 t_1^2}{c^2}}} = \frac{m_0 a_0}{\sqrt{1 - \frac{b_3^2}{c^2}}} \cdot \frac{1 - a_0 b_3 t_1 c^{-2}}{\left(1 - \frac{a_0^2 t_1^2}{c^2}\right)^{\frac{3}{2}}},$$

$$F_{1\beta_1} = \frac{d}{dt_1} \frac{m_0 b_2}{\sqrt{1 - \frac{b_3^2}{c^2}}} = 0,$$

$$F_{1\gamma_1} = \frac{d}{dt_1} \frac{m_0 b_1}{\sqrt{1 - \frac{b_3^2}{c^2}}} = 0.$$

These results are substituted in eqn. (C.83), and, since

$$\frac{d\bar{u}_1}{dt_1} = \frac{d\bar{n}_1}{dt_1} = \frac{d\bar{e}_1}{dt_1} = 0,$$

the following expression is obtained for the force F_1:

$$\bar{F}_1 = \bar{u}_1 \frac{m_0 a_0}{\sqrt{1 - \frac{b_3^2}{c^2}}} \frac{1 - \frac{a_0 b_3 t_1}{c^2}}{\left(1 - \frac{a_0^2 t_1^2}{c^2}\right)^{\frac{3}{2}}} =$$

$$-\bar{k}_1 \frac{m_0 a_0}{\sqrt{1 - \frac{b_3^2}{c^2}}} \frac{1 - \frac{a_0 b_3 t_1}{c^2}}{\left(1 - \frac{a_o^2 t_1^2}{c^2}\right)^{\frac{3}{2}}} .$$

This force, which seems gravitational to the observer in $O_1x_1y_1z_1$, has an expression which depends on the velocity b_3 of the body in the direction in which the force acts, although this velocity was assumed constant in the mobile frame and it is understood that these forces are determined by the presence of a mass, as long as one has to do with a field of such forces.

The result is new compared with that known from classical mechanics, and the approach also differs from that utilized in general relativity.

'It is worth remarking that, for reasonably low velocities for which the approximations $a_0b_3t_1c^{-2} \cong 0$, $a_0^2t_1^2c^{-2} \cong 0$, $b_3^2c^{-2} \cong 0$ hold, the well-known equation $\bar{F}_1 = \bar{u}_1m_0a_0 = - \bar{k}_1m_0a_0$ is obtained; this equation could be obtained immediately by means of classical mechanics. If, in particular, $\bar{F}_1$ is a terrestrial gravitational force and the phenomenon is studied throughout sufficiently narrow intervals for the gravitational acceleration to be considered constant, then $a_0 = g$ (g is the gravitational acceleration in the domain where the phenomenon is studied).

Obviously the procedure holds in the case when the

translation acceleration of frame Oxyz is variable (e.g., according to $a = \lambda\rho^{-2}$ equation where λ is a constant and ρ is the distance between the body M and the fixed pole O_1). More generally, whatever the intricacy of the particular case considered, the inertial and gravitational forces are expressed according to the method presented in the two examples considered. There is one restriction: the relative motion of frames and its influence upon the transport of information (light) have to be such as to allow synchronization of clocks. This fact is to be emphasized.

Note 1. All calculations and reasoning are valid only in frames that allow synchronization of clocks. Unlike the usual approach and interpretation of the theory of relativity, it was assumed that synchronization of clocks is possible not only in inertial frames, but in some of the non-inertial ones as well. More precisely, in those frames in which the law of distribution of relative velocities of the moving spaces (or the gravitational influence) is such that, for every point of one of the two moving spaces, there exists a timing frequency (differing from point to point) in the same equivalence class with the propagation of information taken with respect to the frame in whose space synchronization of clocks is achieved. If this frequency is the same for all the points of the space of a certain frame, then that frame is inertial.

Whenever synchronization of clocks is not possible, use of the methods of general relativity is mandatory.

Note 2. With regard to the gravitational forces, they are known to be determined by the influence of masses. Thus, to obtain their expressions, one has to resort to mobile frames that are Cartesian only within sufficiently narrow space-time intervals.

To expand the study to the whole space, the methods of general relativity have to be utilized. However the method advanced here presents an advantage: it is closer to the classical manner and, in numerous instances, simpler.

C.III. SOME ASPECTS OF CLASSICAL MECHANICS

Classical mechanics is known as an approximation, the result of modelling which, in the study of mechanical motion and in the interpretation of experimental results and findings, does <u>not</u> consider the phenomenon of information transmission, which amounts to assuming that influences (information) propagate instantaneously throughout space. One should add the remaining known components of the model of classical mechanics.

Once this classical concept is accepted, the definition of time given in this book, and implicitly whatever results from the point of view exposed, loses sense. However this concept of classical mechanics represents an approximation, an idealization with multiple consequences, and is, strictly speaking, false (from a scientific point of view). The results of classical mechanics do <u>not</u> comply with the reality recorded by <u>accurate</u> measurements. The approximation is fairly good for the needs of usual technology, when the velocities of phenomena involved are low compared to the velocity of propagation of information.

Therefore, classical mechanics although incorrect in principle, remains a convenient tool and, in many instances, a useful one. For these reasons, classical mechanics will never be abandoned, although, from a rigorous scientific point of view it seems an obsolete discipline.

In the following a modification of the axiomatic basis of the classical model is advanced without abandoning the classical frame.

C.III.1. Grounding mechanics on a single axiom

Classical mechanics may be constructed starting from a single axiom, to rediscover the truths known as the principles of mechanics, as (demonstrated) theorems. This is possible provided that what is known now as the law of conservation of momentum is transformed into a principle.

C.III.1.1. The single axiom of mechanics. In classical mechanics, the mass does not depend on velocity (the relativistic effect of the variation of mass with velocity does not exist) and therefore, when a particle does not capture or emit substance, its mass is constant and equal to the rest mass m, and its momentum (with respect to a given frame) is $\bar{h} = m\bar{v}$ ($\bar{v}$ is the velocity of the particle).

The axiom of classical mechanics may be stated as follows:

The total amount of momentum is constant in the universe, but it may be transferred from one system to another; this phenomenon of momentum exchange between various material systems is referred to as mechanical interaction.

C.III.1.2. <u>Definition of force and derivation of the fundamental laws of mechanics</u>. The variation of momentum of a material system (particles) per unit time is called <u>force</u>.

Thus the force is the rate of momentum variation. If, in the elementary time dt, the momentum of a particle M varies with $d\bar{h}$, then the particle is acted upon by a force

$$\bar{F} = \frac{d\bar{h}}{dt}. \tag{C.87}$$

If particle M interacts with n other material systems, these latter convey to M in a time dt momenta $d\bar{h}_1$, $d\bar{h}_2$, ..., $d\bar{h}_n$, so that

$$\bar{F}_i = \frac{d\bar{h}_i}{dt}; \; i = 1, 2, \ldots, n;$$

$$\bar{F} = \sum_{i=1}^{n} \bar{F}_i = \sum_{i=1}^{n} \frac{d\bar{h}_i}{dt} =$$

$$\frac{\sum_{i=1}^{n} d\bar{h}_i}{dt} = \frac{d\bar{h}}{dt}. \tag{C.88}$$

If, in particular, $d\bar{h} = \sum_{i=1}^{n} d\bar{h}_i = 0$, the particle is in equilibrium.

This is the case of (internal or external) forces which act upon the particles of material systems in equilibrium, a case which is dealt with by statics.

In the case when the force expresses the rate of variation of the momentum exchanged by two material systems, it is called an interaction force (between the two material systems). Obviously such an interaction force may be an active or a binding force.

In the case when the momentum of a material system (a mass point, in particular) varies, but there is no interaction (no momentum exchange with other material systems), the rate of momentum variation is an inertial force called a pseudo-force, or a kinematic force. This force is the result of the behaviour of the frame.

A material system S is called isolated if its particles M_i ($i = 1, 2, \ldots, n$) do not exchange momentum or they exchange only with particles M_j ($j = 1, 2, \ldots, n$) belonging to the same material system S ($M_j \in S$). Thus the momentum of an isolated material system is constant if there are no pseudo-forces.

Consider an isolated material system S consisting of two particles M_1 and M_2 which are not acted upon by pseudo-forces.

The momentum of the whole system is constant:

$$\bar{h} = \bar{h}_1 + \bar{h}_2 = \bar{h}_0 = \text{constant}. \tag{C.89}$$

If the two particles interact with each other and $d\bar{h}_1$ and $d\bar{h}_2$ are the momentum variations of M_1 and M_2, respectively, one can write:

$$\bar{h}_1 + d\bar{h}_1 + \bar{h}_2 + d\bar{h}_2 = \bar{h}_0 = \bar{h}_1 + \bar{h}_2. \tag{C.90}$$

Thus according to (C.89):

$$d\bar{h}_1 + d\bar{h}_2 = 0 \tag{C.91}$$

or

$$\frac{d\bar{h}_1}{dt} + \frac{d\bar{h}_2}{dt} = 0. \tag{C.91'}$$

But:

$$\frac{d\bar{h}_1}{dt} = \bar{F}_{12}, \quad \frac{d\bar{h}_2}{dt} = \bar{F}_{21}, \tag{C.92}$$

where $\bar{F}_{12}$ is the force exerted by M_2 upon M_1 and $\bar{F}_{21}$ the force M_1 exerts upon M_2. It turns out that:

$$\bar{F}_{12} + \bar{F}_{21} = 0, \quad \bar{F}_{12} = - \bar{F}_{21}, \tag{C.93}$$

which is nothing else but the principle of equality of action and reaction, now derived, i.e., transformed into a theorem.

A particle may certainly be subjected to both interaction and inertial forces. If there is no pseudo-force, eqn. (C.87) shows immediately that the particle is isolated, i.e., if $\bar{F} = 0$ then $\bar{h} = m\bar{v} = \bar{h}_0$ = const., and, since $m = m_0$ = const., it turns out that $\bar{v} = \bar{v}_0$ = const., which reconstructs the truth, known as the principle of inertia, in the form of a demonstrated theorem.

Finally, eqn. (C.87) yields $\bar{F} = m \frac{d\bar{v}}{dt} = m\bar{a}$, that is it contains the fundamental law of dynamics which is also given the character of a principle.

Note. For practical applications it is important that, very often, forces may be expressed as functions of space, velocity, and time even before the motion is known.

C.III.2. The inertial or non-inertial character of frames

It is of vital importance, both for theory and practice, to estimate correctly the affordable approximation related to the inertial or non-inertial character of a frame.

C.III.2.1. <u>Inertial frames</u>. These are frames in which inertial forces do not manifest themselves. Therefore, whenever the motion of an isolated material system with respect to an inertial frame is studied, the momentum is constant:

$$\bar{H} = m\bar{v}_C = \bar{H}_0 = \text{constant}, \tag{C.94}$$

where m is the total mass of the material system and $\bar{v}_C$ is the velocity of its center of mass.

Eqn. (C.94) means in fact that $\bar{v}_C = \bar{v}_{0C}$ = const. since m is assumed not to change.

It follows that a frame is inertial if the motion of the centre of mass of a material system with respect to it is rectilinear and uniform.

However, there are no <u>completely</u> isolated systems. Therefore, the defining of an inertial frame assumes a certain degree of approximation corresponding to the assimilation of a material system (whose interaction with the surrounding is relatively weak) to an isolated system. Naturally, the weaker the interaction of system S with other material systems, the closer to the ideal concept of inertial frame are the frames with respect to which the centre of mass of S moves rectilinearly and uniformly.

E.g., a frame with respect to which the centre of the Earth is in uniform rectilinear motion (or at rest) is, to a fairly rough approximation, inertial.

Much closer to the ideal concept of inertial frame is a frame in which the mass centre of the solar system moves rectilinearly and uniformly.

One can get even closer to this concept if the centre of mass of the Galaxy is considered. Obviously all this is so due to the various degrees of mechanical isolation.

C.III.2.2. Non-inertial frames. It should be recalled that all points in space which are at rest with respect to a frame make up the space of that frame.

Consider two frames R_0 and R, and let σ_0 and σ be their spaces. This means that any points $M_0 \in \sigma_0$ and $M \in \sigma$ are at rest with respect to R_0 and R, respectively.

If R_0 is an inertial frame and any point $M \in \sigma$ is in uniform and rectilinear motion with respect to R_0, then R is an inertial frame.

If, however, $M \in \sigma$ is in accelerated motion with respect to R_0 (i.e., this motion is neither uniform nor rectilinear), then R is a non-inertial frame.

Note. Regarding the behaviour of light in inertial or non-inertial frames and the possible influence of masses upon the propagation of light beams, both phenomena are irrelevant for classical mechanics, since the transmission of information is not considered in the reasoning and calculations of Newtonian mechanics.

This has made possible the remarkable simplification in the definition of inertial and non-inertial frames in the sense adopted here.

INDEX